Immersive Multimodal Interactive Presence

Springer Series on Touch and Haptic Systems

Series Editors

Manuel Ferre
Marc O. Ernst
Alan Wing

Series Editorial Board

Carlo A. Avizzano
José M. Azorín
Soledad Ballesteros
Massimo Bergamasco
Antonio Bicchi
Martin Buss
Jan van Erp
Matthias Harders
William S. Harwin
Vincent Hayward
Juan M. Ibarra
Astrid Kappers
Abderrahmane Kheddar
Miguel A. Otaduy
Angelika Peer
Jerome Perret
Jean-Louis Thonnard

For further volumes:
www.springer.com/series/8786

Angelika Peer • Christos D. Giachritsis

Editors

Immersive Multimodal Interactive Presence

 Springer

Editors
Angelika Peer
Institute of Automatic Control Engineering
Technische Universität München
Munich, Germany

Christos D. Giachritsis
Research and Development
BMT Group Ltd.
Teddington, UK

ISSN 2192-2977   e-ISSN 2192-2985
Springer Series on Touch and Haptic Systems
ISBN 978-1-4471-6213-1   ISBN 978-1-4471-2754-3 (eBook)
DOI 10.1007/978-1-4471-2754-3
Springer London Dordrecht Heidelberg New York

British Library Cataloguing in Publication Data
A catalogue record for this book is available from the British Library

Springer is part of Springer Science+Business Media (www.springer.com)

Series Editors' Foreword

This is the fourth volume of the "Springer Series on Touch and Haptic Systems", which is published in collaboration between **Springer** and the **EuroHaptics Society**. *Immersive Multimodal Interactive Presence* is focused on exploring how the haptic component increases multimodal human interactions. Three paradigms of human interaction are considered: a person manipulating a virtual object, two people directly interacting, and two people interacting by using a shared object. These studies were carried out under an Integrated Project called Immersence, which was funded by the 6th Framework Program of the European Union. These kinds of projects usually generate a lot of conference and journal papers and, normally, this knowledge is disseminated without the opportunity of compiling the information properly.

The editors have gone to considerable lengths in *Immersive Multimodal Interactive Presence* to compile the most relevant results and organize them in a comprehensive manner. Moreover, two additional chapters have been included in order to bring in two further important issues related to Immersence.

The series editors are confident in the interest that this publication will create and would like to invite other consortiums working on haptics to prepare similar publications. These books provide an excellent opportunity to overview the developing state of the art that other researchers will find invaluable as groundwork for their own future studies in this important research area.

Manuel Ferre
Marc Ernst
Alan Wing

Preface

The addition of haptic feedback to virtual environments has enhanced users' virtual experience dramatically. It has enabled users to experience physical characteristics of objects such as weight, temperature, roughness and compliance, which was previously impossible with only visual and auditory feedback. Moreover, haptic feedback has the potential of greatly enhancing the experience of social interactions within virtual environments.

This book summarizes results achieved in the IMMERSENCE research project funded by the 6th Framework Programme of the European Union, FET—Presence Initiative. In addition, it includes selected chapters of other researchers in the field who were not part of the Immersence consortium. Different aspects of haptic interaction are explored based on three different scenarios: Person-Object (PO), Person-Object-Person (POP) and Person-Person (PP) interaction. While the PO scenario focuses on multimodal information processing, manipulative aspects of haptic interaction, and subliminal perception, the POP scenario explores examples of joint action in virtual environments, such as collaborative lifting. Finally, the PP scenario uses paradigms of handshaking and dancing to investigate direct contact between a human and an agent located in virtual reality.

The book is an example of interdisciplinary research directed towards the realization of multimodal immersive virtual environments with particular focus on haptic interaction. Recent results of psychophysical and behavioral studies are reported along with new technological developments for haptic displays and novel haptic rendering techniques.

The book is organized in two parts: I. Psychophysical and Behavioral Basis and II. Technology and Rendering. Chapters in part I report studies investigating important aspects of individual and collaborative object manipulation. Studies exploring the PO scenario investigate conditions affecting tactile sensitivity during manipulation. Other studies evaluate the effectiveness of multipoint contact haptic interfaces in simulating weight during unimanual and bimanual manipulation of virtual objects and examine the role of grip forces and cutaneous feedback on weight perception. Studies on multisensory enhancement, emotional factors and subliminal cues on performance in virtual environments are also reported. The POP studies are focusing on experiments investigating human-human and human-robot collaborative lifting and suggest control algorithms describing this type of collaborative behaviour. The part also includes a chapter on issues related to the use of psychological experiments to evaluate artificial haptic interaction partners.

Part II reports on advanced rendering techniques for data-driven visuo-haptic rendering of deformables objects, haptic rendering for multiple contact points, as well as the rendering of cutaneous cues for the PO scenario. For the PP scenario rendering of social haptic interaction partners is investigated and a new MR compatible sensing glove to analyze dyadic interactions using fMRI is developed. Finally, the realization of advanced human-computer collaboration schemes involving haptic role exchange and negotiation mechanisms are investigated for the POP scenario.

Although the two parts of the book are virtually separated, they are not in practice: both of them benefit from each other and many overlappings between the two parts exist. Novel devices and rendering techniques cannot be developed without the knowledge of psychophysical and behavioral findings, while new experiments cannot be performed without sophisticated devices and rendering techniques. Consequently, the two parts are highly complementary and many of the synergies between them are highlighted in the single chapters.

Finally, we would like to thank all the authors for their valuable contribution to this book.

Germany   Angelika Peer
UK   Christos D. Giachritsis

Contents

Contributors

Jorge Barrio Centre for Automation and Robotics UPM-CSIC, Universidad Politecnica de Madrid, Madrid, Spain

Cagatay Basdogan Koc University, Istanbul, Turkey

Matteo Bianchi Interdepartmental Research Center "E. Piaggio", University of Pisa, Pisa, Italy

Antonio Bicchi Interdepartmental Research Center "E. Piaggio", University of Pisa, Pisa, Italy

Daniela Bonino Laboratory of Clinical Biochemistry and Molecular Biology, University of Pisa Medical School, Pisa, Italy

Martin Buss Institute of Automatic Control Engineering, Technische Universität München, München, Germany; Institute of Automatic Control Engineering, Technische Universität München, Munich, Germany

Pablo Cerrada Centre for Automation and Robotics UPM-CSIC, Universidad Politecnica de Madrid, Madrid, Spain

Danilo De Rossi Interdepartmental Research Center "E. Piaggio", University of Pisa, Pisa, Italy

Satoshi Endo Behavioural Brain Sciences Centre, University of Birmingham, Birmingham, UK

Marc O. Ernst Department of Cognitive Neuroscience, University of Bielefeld, Bielefeld, Germany

Paul Evrard Laboratoire d'Intégration des Systémes et des Technologies (LIST), Commissariat à l'énergie atomique et aux énergies alternatives (CEA), Fontenay-aux-Roses, France

Manuel Ferre Centre for Automation and Robotics UPM-CSIC, Universidad Politecnica de Madrid, Madrid, Spain

Daniela Feth Institute of Automatic Control Engineering, Technische Universität München, Munich, Germany

Michael Fritschi Department of Biomedical Engineering, Khalifa University, Abu Dhabi, UAE

Christos D. Giachritsis Research and Development, BMT Group Ltd., London, UK

Raphaela Groten Institute of Automatic Control Engineering, Technische Universität München, Munich, Germany

Matthias Harders Computer Vision Lab, ETH Zurich, Zurich, Switzerland

Valentina Hartwig Institute of Clinical Physiology, National Research Council, Pisa, Italy; Interdepartmental Research Center "E. Piaggio", University of Pisa, Pisa, Italy

Raphael Hoever Computer Vision Lab, ETH Zurich, Zurich, Switzerland

Jens Hölldampf Institute of Automatic Control Engineering, Technische Universität München, München, Germany

Abderrahmane Kheddar CNRS-AIST Joint Robotics Laboratory (JRL), UMI3218/CRT, Tsukuba, Japan; CNRS-UM2 LIRMM, Interactive Digital Human, Montpellier, France

Ayse Kucukyilmaz Koc University, Istanbul, Turkey

Luigi Landini Department of Information Engineering, University of Pisa, Pisa, Italy

Salih Ozgur Oguz Koc University, Istanbul, Turkey

Angelika Peer Institute of Automatic Control Engineering, Technische Universität München, München, Germany

Serge Pfeifer Computer Vision Lab, ETH Zurich, Zurich, Switzerland

Pietro Pietrini Laboratory of Clinical Biochemistry and Molecular Biology, University of Pisa Medical School, Pisa, Italy

Kyle B. Reed University of South Florida, Tampa, FL, USA

Miriam Reiner Department of Education in Technology and Science, Technion—Israel Institute of Technology, Haifa, Israel

Emiliano Ricciardi Laboratory of Clinical Biochemistry and Molecular Biology, University of Pisa Medical School, Pisa, Italy

Enzo Pasquale Scilingo Interdepartmental Research Center "E. Piaggio", University of Pisa, Pisa, Italy

Tevfik Metin Sezgin Koc University, Istanbul, Turkey

Alessandro Tognetti Interdepartmental Research Center "E. Piaggio", University of Pisa, Pisa, Italy

Nicola Vanello Department of Information Engineering, University of Pisa, Pisa, Italy

Marco P. Vitello Multisensory Perception and Action Group, Max Planck Institute for Biological Cybernetics, Tübingen, Germany

Zheng Wang Institute of Automatic Control Engineering, Technische Universität München, München, Germany

Thibaut Weise Computer Vision Lab, ETH Zurich, Zurich, Switzerland

Alan M. Wing Behavioural Brain Sciences Centre, University of Birmingham, Birmingham, UK; School of Psychology, University of Birmingham, Birmingham, UK

Raul Wirz Centre for Automation and Robotics UPM-CSIC, Universidad Politecnica de Madrid, Madrid, Spain

Chapter 1
Introduction

Christos D. Giachritsis and Angelika Peer

Abstract Research on high-fidelity visual and audio feedback has dominated the design and development of virtual environments for nearly half a century. Over the last decade the importance of haptic feedback has been recognized and extensive effort has been devoted to haptic interaction research. This book explores haptic feedback in virtual environments based on three different scenarios involving Person-Object (PO), Person-Object-Person (POP) and Person-Person (PP) interactions. The book highlights the interdisciplinary nature of haptic interaction research and reports recent results of psychophysical and behavioral studies along with new technological developments for haptic displays and novel haptic rendering techniques. In this introductory chapter, we provide motivation for this interdisciplinary approach followed by a detailed description of chapters composing this book.

While the importance of high-fidelity visual and audio feedback was recognized very early in the design of virtual environments, haptic interaction was mainly ignored or restricted to input devices such as mouse and joystick. Over the last decade considerable research and commercial efforts have been committed into emphasizing the importance of haptic interaction in improving the effectiveness and realism of virtual environments. As a result, new haptic interfaces and haptic rendering algorithms have been developed, which have transformed the way we perceive virtual environments and interact with virtual objects and characters.

Today is possible to use haptic interfaces to perceive a wealth of haptic properties of virtual objects such as size, shape, texture, compliance and weight. This has been possible through continuous improvement of haptic devices based on new actuation principles as well as advanced haptic rendering algorithms. Nonetheless, these advances would not be possible without the collaboration of researchers work-

C.D. Giachritsis (✉)
Research and Development, BMT Group Ltd, Goodrich House, Waldegrave Road 1,
TW11 8LZ Teddington, UK
e-mail: cgiachritsis@bmtmail.com

A. Peer
Institute of Automatic Control Engineering, Technische Universität München, Munich, Germany
e-mail: angelika.peer@tum.de

A. Peer, C.D. Giachritsis (eds.), *Immersive Multimodal Interactive Presence*,
Springer Series on Touch and Haptic Systems,
DOI 10.1007/978-1-4471-2754-3_1, © Springer-Verlag London Limited 2012

ing in different disciplines including engineering, neuroscience, computer science and psychology. Research into how the human haptic system works has provided designers and developers of haptic interfaces with valuable information on how to stimulate human touch effectively to achieve more realistic multimodal interactions with virtual environments. Moreover, technological advances in haptic devices have provided researchers in human haptics with valuable tools to conduct novel experiments which would be impossible without these devices.

This book presents work based on this interdisciplinary approach by connecting psychophysical and behavioral haptics research to advances in haptic technology and haptic rendering. Three different scenarios representing different types of interactions in virtual environments are explored: Person-Object (PO), Person-Object-Person (POP) and Person-Person (PP) interaction. The PO scenario concentrates on the interaction of a single person with a passive real and virtual object. The POP scenario considers co-manipulation of real and virtual objects between two people, and between a person and a robot. The PP scenario considers direct interactions between two partners (human-human, human-robot, or human-agent) who are involved in social activities such as handshaking and dancing.

The book is a collection of work in the field covering a series of possible interaction types that are summarized in the three aforementioned scenarios.

The first part of the book, presents experimental work in human haptics with particular emphasis on behavioral and psychophysical aspects of PO, POP and PP interactions.

Chapter 2 reports experiments investigating conditions of *tactile suppression*. The authors suggest that tactile sensitivity is affected by active movements due to pre-motor command and sensory input which may interfere with the tactile input.

Chapter 3 investigates the issue of *weight perception in real and virtual environments*. Unimanual and bimanual weight perception during manipulation of virtual objects is evaluated. The results show that it is feasible to achieve affective simulation of bimanual weight but the sensitivity to weight changes decreases dramatically. The possible roles of bimanual grip forces, and the contributions of proprioceptive and cutaneous cues are also addressed.

Chapter 4 investigates the *cognitive processes that lead to the construction of meaning* through the experiencing of dynamic haptic patterns. The author explores the role of subliminal cues in the formation of a haptic language and shows that even though participants are not aware of these cues their perception and action are affected by them.

Chapter 5 addresses important issues when using *psychological experiments to design and evaluate artificial partners* (e.g., robots, avatars) that can perform collaborative tasks with human partners. The authors introduce the concept and role of psychological experiments and layout the challenges that researchers face when using them during the design process. These issues are illustrated through two examples of applying psychological experimentation in early and late design stages. The first examines the rise of dominant partners in a POP task and the second, evaluates different types of partners.

Chapter 6 examines the role of non-verbal (e.g., tactile) cues in *coordinated action between two partners* (e.g., human-human and human-robot). The authors review three studies investigating the predictive and reactive control of motor behavior during collaborative lifting (POP scenario). A model is suggested in which partners adjust their motor behavior on the basis of the outcome of the previous trial.

The haptic interaction of two partners is also the focus of Chap. 7 which *compares the accuracy of individual and collaborative action*. The reported studies suggest that through the application of force feedback the two partners could complete a task more accurately than they would have done if they performed the same task individually.

In terms of technology and algorithms for rendering different scenes and interactions, a series of advances are described that help making virtual environments much more realistic than experienced today and thus, bring us closer to the ultimate goal of total immersion into virtually rendered scenes. For the PO scenario a series of highly-sophisticated algorithms for visuo-haptic rendering of objects are presented.

In Chap. 8 *visuo-haptic rendering of deformable bodies* is studied using a data-driven record-replay-recreate method, which contrasts classical methods based on physical or heuristic models. The overall visuo-haptic acquisition and rendering system is introduced along with a detailed description of the visual and haptic recording mechanisms and rendering algorithms.

Chapter 9 advances the state of the art by focusing on *haptic rendering methods for multiple contact points* as required for example when grasping or manipulating virtual objects. A software architecture for the haptic rendering of multiple contact points is introduced and experiments for bimanual and cooperative manipulation are described and analyzed.

Simultaneous rendering of kinesthetic and tactile information is investigated in Chap. 10. The combination of a commercially kinesthetic haptic display combined with a newly designed tactile flow-based display is presented that is able to simultaneously render force/area and force/displacement relationships. For spatial rendering of tactile cues a pin-based mechanical stimulator is evaluated for its capability of rendering simple shapes and moving patterns.

The rest of the book deals with direct and object-mediated human-human interactions. Chapter 11 concentrates on the PP scenario by looking at *social interactions with virtual agents*. Taking handshaking and dancing as prototypical examples, methods for rendering interactive behavior to estimate and communicate intentions as well as to adapt the behavior to the interaction partner are investigated following a record-replay-recreate approach.

Chapter 12 also focuses on the PP scenario and presents new technological developments that allow performing interaction studies in MR environments. A new *fMRI compatible sensing glove* and its calibration procedure are presented along with obtained results of MR compatibility. A preliminary study on differences of neural correlates of human-human and human-robot handshaking shows the potential of the newly developed technology.

Finally, the POP scenario is investigated in Chap. 13 that focuses on human-computer cooperation and implements *dynamic role exchange and haptic negotiation mechanisms*. Two cooperative games are realized and analyzed to demonstrate the benefit of the newly defined role-exchange mechanisms and haptic negotiation framework. Further, the effectiveness of visual and haptic cues in conveying negotiation-related states are investigated.

The three presented scenarios have also several commercial applications. Applications of the PO scenario span from advanced teleconferencing systems that allow transferring objects from one place to another, over e-commerce systems that allow experiencing products even before they are purchased, to interactive computer-aided design systems that support collaborative product development. Applications for the PP and POP scenario range from physical robot assistants that operate in industrial settings and help in moving large and heavy objects, over walking assistants, devices for motor-rehabilitation and motor skill learning, to virtual computer assistants that support humans in performing tasks at a distance using teleoperation technology and social haptic interaction partners that interact socially, e.g. in performing a handshake.

Part I
Psychophysical and Behavioral Basis

Chapter 2
Active Movement Reduces the Tactile Discrimination Performance

Marco P. Vitello, Michael Fritschi, and Marc O. Ernst

Abstract Self-performed arm movements are known to increase the tactile detection threshold (i.e., decrease of tactile sensitivity) which is part of a phenomenon called tactile suppression. Today, the origin and the effects of tactile suppression are not fully understood. Tactile discrimination tasks have been utilized to quantify the changes in tactile sensitivity due to arm movements and to identify the origin of tactile suppression. The results show that active arm movement also increases tactile discrimination thresholds which has never been shown before. Furthermore, it is shown that tactile sensitivity drops at approximately 100 ms before the actual arm movement. We conclude that tactile suppression has two origins: (1) a movement related motor command which is a neuronal signal that can be measured 100 ms before a muscle contraction. This motor command is the origin for the increase of the discrimination threshold prior to the arm movement and (2) task irrelevant sensory input which reduces tactile sensitivity after the onset of the arm movement.

2.1 Introduction

One of the brain's major challenges is to filter task relevant from task irrelevant information. Mechanisms which filter out irrelevant information are suggested to exist across many sensory systems [1] and the perceptual consequences are in some cases well understood.

M.P. Vitello (✉)
Multisensory Perception and Action Group, Max Planck Institute for Biological Cybernetics,
Tübingen, Germany
e-mail: MarcoVitello@web.de

M. Fritschi
Department of Biomedical Engineering, Khalifa University, Abu Dhabi, UAE
e-mail: michael.fritschi@kustar.ac.ae

M.O. Ernst
Department of Cognitive Neuroscience, University of Bielefeld, Bielefeld, Germany
e-mail: marc.ernst@uni-bielefeld.de

A. Peer, C.D. Giachritsis (eds.), *Immersive Multimodal Interactive Presence*,
Springer Series on Touch and Haptic Systems,
DOI 10.1007/978-1-4471-2754-3_2, © Springer-Verlag London Limited 2012

Tactile suppression, a decrease in tactile sensitivity during active limb movement, seems to be a phenomenon that results from some form of filtering. In everyday life, the sense of touch has to deal with a vast amount of sensory information. One can speculate that tactile information may be divided into two groups of information types in order to keep the brain's workload low: task relevant and task irrelevant. During normal body movements, for example, many sensory inputs are generated by one's clothing rubbing on the skin, air flow due to limb movements and others. These tactile stimuli are usually perceptually unimportant for our behavior and most of the time they remain unconscious. Here, tactile suppression seems to reduce tactile sensitivity in order to keep tactile stimulation, which results from self performed movement, low.

Intuitively, one would expect the brain to filter out irrelevant stimuli in order not to be involved in processing information which is irrelevant for a specific task. Hypothetically, two kinds of information filters may be possible:

(1) Active filters, whereby the neural system is actively preprocessing information and then separating relevant from irrelevant information. An active filter needs processing power. This processing power should be lower than that which would be needed to process the irrelevant information in order to lower the brain's workload. As an analogy one can see it as a spam filter where messages which contain suspicious keywords are filtered out before they will be sent from one server to the other through the World Wide Web.

(2) Passive filters can filter out information using a fixed criterion. Passive filters are suggested to be part of the peripheral nervous system or early stages of the central neural system. In this case, only little processing capacity is necessary, if at all. As an example one could see it as an instance that filters out messages that have only a small file size, without reading the content of the message.

Filter mechanisms have already been identified, for example, in the visual system. Saccadic suppression is a phenomenon whereby visual sensitivity is decreased during ballistic eye movements (so-called saccades) and may be seen as a visual analogue to tactile suppression, because both phenomena reduce the amount of sensory information during active movement [2–4]. Saccades generate rapid image changes on the retina, which are not of importance, comparable to the tactile stimuli (e.g., cloth rubbing on the skin) caused by limb movements. Saccadic suppression can easily be demonstrated if an observer is standing in front of a mirror. If he is looking into his left and subsequently into his right eye, he will not perceive any eye movement. To explain this phenomenon it has to be assumed that the sensitivity of the visual system must be decreased during saccades, which was already claimed by Helmholtz (1867) [5]. A mechanism which explains the functioning of saccadic suppression is the reafference principle [6], in which a copy of the motor command (efference copy) is compared with the afferent signal (reafference) that comes from an effector (e.g., muscle). Normally, both signals cancel each other out on the level of the center 1 (Fig. 2.1). If the signal strength of the afference becomes too small or too big due to an external influence, then a proportion of the afferent signal remains (exafference). This is the signal for object motion.

Fig. 2.1 Schematic illustration of the reafference principle. With kind permission from Springer Science+Business Media: [6], Fig. 2.4. A center may be seen in this context as a general level of information processing

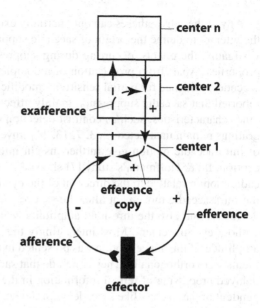

It is still unknown whether both, tactile and visual suppression, share only the sensory consequences, i.e., decrease of sensory abilities during body movements, or whether the decrease in sensitivity is due to common underlying processes where body movement interacts with perception in the same way independent of the sensory modality. The phenomenon of sensory suppression will be explained by means of the saccadic suppression which is well understood [7, 8].

In previous studies, effects of saccadic suppression, i.e., the decrease of visual sensitivity, were identified and underlying mechanisms have been proposed as it is shown in the following brief overview:

Sperry (1950) [9] suggested that corollary discharges of motor patterns into the sensory system may be crucial for the adjustment of a visually stable world during saccades. This theory is a variation of the reafference principle.

Holt (1903) [10], creator of the inflow theory, claims signals from proprioceptors of the extraocular muscles to be responsible for a blocking of visual centers. Thus, according to Holt the visual system is calibrated by sensory signals and not by eye movement motor commands.

Richards (1969) [4] proposed that shear forces, during saccades, between the retina and the vitreal body induce a blocking of the information stream through the optic nerve. This may explain the drop in visual sensitivity but not the adjustment of a stable visual world.

Matin (1974) [8] suggested that the time in which the photoreceptors are exposed to light (i.e., a visual stimulus) is too short, due to the high motion velocity of the eyes during saccades. The photoreceptors would not rise above threshold.

Finally, Mackay (1970) [11] has shown that rapid image motion itself (as it occurs during saccades) can induce a suppressing effect, while the eyes remain stationary.

Any of these hypotheses can only partially explain saccadic suppression. Despite the attempt to define the origin of saccadic suppression, many researchers were investigating the effects, occurring during suppression, and the resulting perceptual properties. Apart from propagation of a complete loss of visual sensitivity during saccades, evidence for partial sensitivity modification is found. For example, it was reported that saccadic suppression mainly affects motion perception [12], whereas other characteristics, such as contrast detection of high-frequency or equiluminant gratings remain unaffected [2, 3, 7, 13, 14]. Investigations regarding the perception of intrasaccadic motion give further insight into the directional sensitivity of perception. Ilg & Hoffmann's study [7] showed by changing the direction of intrasaccadic motion relative to the direction of the eye movements that subjects perceived the intrasaccadic movement after the saccade. The perceived motion velocity was decreased and also the threshold amplitude was increased compared to perception without eye movement. Most interestingly they did not find changes in threshold amplitudes if the image motion was presented in the same direction as the eye movements were orthogonally. They conclude that saccadic suppression is mediated by a delayed processing of retinal information in the CNS during saccades and is independent of parameters like saccade amplitude or eye motion direction.

To identify whether saccadic suppression results from a central signal as it is proposed by [9] or [6], or whether visual motion caused by the eye movement itself is inducing the visual masking effect during saccades is a major question. Researchers have found some indication that both are acting on the same cortical mechanisms. An interesting finding, that supports the idea of a central mechanism, is the early action of saccadic suppression, that is the finding that suppression precedes visual motion analysis [12] and contrast masking [3]. The time course for the loss of sensitivity has been shown to begin as early as 50 ms before the onset of saccades, has the highest sensitivity reduction at or shortly before the start of saccades, diminishes during saccades, and vanishes after the end of saccades [15].

The tactile sense shows also sensitivity changes during self-performed movements. Referring to the visual phenomenon it is called tactile suppression or tactile gating whereby tactile stimulation during self-produced body movement is felt less intensely than stimulation during not self-produced movements. A well-known version of this filtering effect is that you cannot tickle yourself [16–20] or at least the sensation of self-tickling is less powerful. This can be interpreted to serve as a strategy to reduce the brain's workload by keeping out uninformative stimuli. As it is supposed in other sensory systems, separating sensory information due to its task relevance can reduce the workload of the brain and is supposed to be a major goal of tactile gating [21].

How tactile suppression is generated is not fully understood, yet. However, two major mechanisms are thought to be involved in decreasing tactile sensitivity during self-motion. Firstly, as in saccadic suppression, the reafference principle [6] is claimed to mediate tactile suppression [22]. Sensory consequences of self-performed movements can be predicted precisely [16, 20] and can therefore be cancelled out. Unpredicted stimuli will not be cancelled and become accentuated. Secondly, it has been shown that the motor command, which is responsible for the execution of muscle contraction, induces a blockade between the peripheral- (PNS) and the central nervous system (CNS) [23]. Thus, afferent sensory information is filtered out due to this sensory gating.

Similar to saccadic suppression, it is reported for tactile suppression to be selective for decreasing the sensitivity only for a subset of tactile features while others seem to be unaffected. Most studies report on an increased detection threshold during active movement [22, 24–28], while the discrimination threshold is stated to remain unchanged [24, 25, 29].

In most of the studies on tactile gating, changes in detection and discrimination thresholds have been investigated by applying electrical stimuli to the skin or by directly stimulating nerves. Intensity ratings were utilized to determine the thresholds. However, this method has two major disadvantages: (1) Electrical stimulation is not common in real life. This stimulation is hardly comparable to tactile sensing as it is experienced during tactile exploration. Manipulative finger movements, for example, generate multiple kinds of stimuli such as normal force, shear force or vibration. Electrical stimulation however does not stimulate mechanoreceptors specifically. (2) Electrical stimulation provides the possibility to investigate intensity discrimination but it does not permit for directional changes of moving stimuli. Moving stimuli for example, which occur during any tactile exploration are hardly possible to be simulated by electrical stimulation.

Previous findings, that tactile detection threshold changes during active movement but discrimination threshold does not, may be an exceptional case of intensity ratings of electrical stimulation.

To investigate whether active movement affects the tactile discrimination threshold, real life tactile stimuli which occur during tactile exploration are utilized (i.e., normal and tangential forces). Normal forces can be generated if one is taping on a rough surface and the skin is indented perpendicular to the skin surface. In this study, normal forces will be generated by a set of Braille generators, consisting of pins which can be extended vertically. Tangential forces (i.e., forces parallel to the skin surface) appear during almost any tactile manipulation, for example if one is rubbing the finger on a surface. In this study they are generated with a tangential force device, consisting of a pin which is able to move on a horizontal x/y-plane. The fact that it can be changed in intensity but also in direction makes it a good stimulus to investigate the effect of active movement on tactile sensitivity of moving stimuli. This provides the possibility not only to study the effect of movement on tactile sensitivity but also a basic phenomenon of motion perception, namely, the involvement of movement prediction on the amount of sensitivity decrease. Considering the reafference principle, described above, it seems plausible that the arm movement direction may be taken into account to modulate tactile sensitivity in a direction specific manner. Thus, tactile stimulus motion along the same direction of the arm movement may be attenuated because the sensory consequences of a self performed arm movement are predicted and canceled out [16–18]. Tactile motion perpendicular to the arm movement may be less predictable, because they hardly ever appear in a real environment and thus, the sensory system may have been hardly ever confronted with perpendicular tactile movement. The unpredictability of these stimuli may increase the perceptibility which would result in a higher sensitivity for stimuli which move perpendicularly to the arm movement.

Tactile gating, however, suggest a general decrease in sensitivity since it is simply blocking the information transfer to the central nervous system, caused by a motor command. Consequently, any sensory information will be blocked which results in a lower amplitude of sensory signals in the somatosensory cortex leading to a lower quality of the tactile stimulus.

The aim of this study is to measure qualitatively and quantitatively tactile sensitivity changes during arm movement under real life conditions. That is, utilizing realistic tactile stimuli combined with natural arm movements. Furthermore, the predictability of tactile stimuli and its effect on the directional sensitivity changes will be investigated.

2.2 Experiment 1: Sensitivity Reduction vs. Predictability of Arm Movement

The aim of this experiment was to investigate whether tactile sensitivity changes for different tactile motion directions relative to the direction of an arm movement. This was done by simulating a tactile exploration scenario in which one is sliding the index finger over a surface. Tactile motion, generated by a pin which moves parallel to the skin surface, was applied to the fingertip. Such tangentially moving stimuli are present during most tactile manipulations and were combined with a simple forward movement of the arm. The tactile stimuli can be changed in their motion direction so that they move in line with the direction of arm motion or with an angular offset. By measuring the discrimination threshold for deviations of the tactile motion direction during or in the absence of arm movements it is possible to investigate the following question:

Is there a change in the tactile discrimination threshold during active arm movements or will it be unaffected as it is claimed in previous studies. Today, the prevailing view is that active movement increases the tactile detection threshold but not the discrimination threshold.

If the discrimination threshold changes as a result of active arm movements, two hypotheses will be tested regarding the sensory consequences of tactile suppression:

(1) If the prediction of sensory consequences is taken to cancel out self performed sensory input and thus, to alter tactile sensitivity during active movement, then differences in the discrimination threshold during self-performed arm movements should occur. If the motion direction of the tactile stimulus and the arm movement are congruent (i.e. tactile consequences are predictable) the discrimination threshold should be higher than if both stimuli move in orthogonal directions (i.e. tactile consequences are not predictable).

(2) If tactile gating is a major factor that causes tactile suppression, then no difference in the discrimination threshold should occur. The gating effect is expected

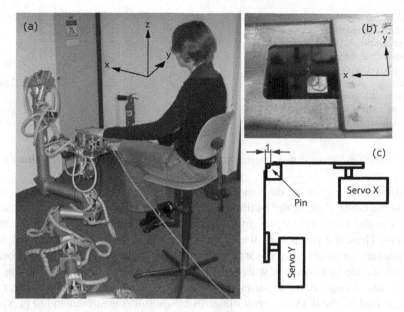

Fig. 2.2 (**a**) Participant seated in front of the kinesthetic (ViSHaRD 10) and the tactile device. (**b**) Close-up view of the pin of the tactile motion device. (**c**) Schematic diagram (top view) of the tactile motion device. Two servo motors are generating the pin motion in x and y direction. Pin diameter is 1 mm. Pin is at the level of the cover plate

to reduce any tactile input, regardless of the motion direction of the arm and the tactile stimulus.

2.2.1 Methods

Twelve right-handed, participants (seven male, five female) participated for pay (8€/h). Ten of them were naïve to the purpose of the experiment. Participant age ranged from 19 to 35 years (average 26 years). None of them reported previous injuries or impairment of tactile sensitivity of the finger tip. The experiments were approved by the local ethics committee.

A custom made hyper-redundant kinesthetic display, called ViSHaRD 10 (for details see [30]), was used to provide kinesthetic force feedback to the operator's hand and to track its position (Fig. 2.2). As a tactile stimulator a custom made tactile motion device was utilized to generate tactile stimuli that move parallel to the skin (for details see [31]). A metal pin moved radially on the x/y plane to exert tactile motion stimuli. The radius of tactile motion is 1 mm. The maximal pin velocity is 10 mm/s. Figure 2.2 shows the tactile device in use (a) with a close up view of the pin area (b) and a schematic diagram of the top view of the pin (c).

The tactile motion device was mounted on the kinesthetic device in order to couple kinesthetic and tactile stimulation. For our experiment the free moving space

Fig. 2.3 Possible directions of the standard and the comparison stimuli. The *left panel* shows the three standard directions (0°, 45°, 90°); the *right panel* shows the possible comparison directions relative to the standard directions

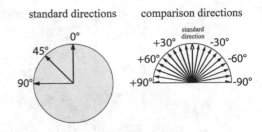

of the kinesthetic device was limited to one degree of freedom (1 DOF) along the *x* axis.

Participants were sitting on a bar stool in front of the setup. Their left hand was fixed to the cover plate of the tactile device. Adhesive tape was used to ensure the fixation of the hand and to prevent unintentional hand movements during the experiment. Thus, the position of the index finger was kept constant during all the experimental conditions. Participants were instructed to move the arm of the force feedback device by pushing it with the thenar eminence in order to keep the finger free of force throughout the experiment. The distal phalanx of the left index finger is not supported by the device's cover plate. To ensure full skin stretch and to prevent that the pin slides over the finger tip, the metal pin which mediated the skin stretch was glued with a minimal amount of cyanoacrylate to the tip of the index finger. Sight of the index finger was prevented by cardboard blinds. The lateral surfaces of the tactile device were covered with cardboard to prevent sight of the device's mechanics.

White noise presented via headphones effectively masked the noise caused by the mechanics of the kinesthetic and tactile devices. Two IBM-compatible PCs (one for the tactile and one for the kinesthetic device) controlled the stimulus presentation, data collection and pin motion of the tactile device, using custom programmed applications.

The tactile stimulus was a pin stroke generated by the tactile motion device. A single stroke consisted of a tactile motion in the forward and backward direction of one metal pin, starting and ending at a center position of the tactile device. The amplitude of the pin movement was 1 mm in length for each direction with a velocity of 10 mm/s resulting in a stroke presentation time of 200 ms. Stroke directions are characterized by their radial deflection as illustrated in Fig. 2.3.

Three standard orientations were chosen for the tactile strokes, i.e., 0° (towards the finger tip) 45° and 90°. Each standard orientation was paired with a set of 19 comparison strokes ranging from ±90° around the standard direction in steps of 10° (Fig. 2.3). An 84%-discrimination threshold was measured for every standard direction using the method of constant stimuli in a two-interval forced-choice paradigm. The discrimination task was performed in three experimental conditions, which were "static", "active" and "passive" (see below). Each of the three experimental conditions comprised twelve repetitions of each standard and its 19 comparison strokes (the order of the intervals in which the pairs of strokes were presented was randomized) resulting in 228 trials per condition. All the trials were presented in three blocks of 228 trials each. Each block lasted about 20 minutes. Between each

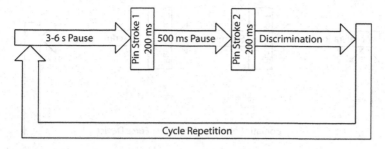

Fig. 2.4 Schematic flow chart of a trial in the "static" condition

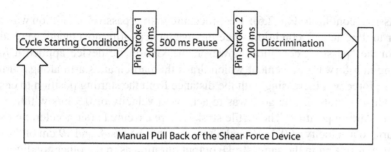

Fig. 2.5 Schematic flow chart of a trial in the "active" condition

block there was a 5 minute break. So collecting data for one condition lasted a bit more than an hour per subject. All three conditions were tested for all subjects on different days. The order of conditions (i.e., static, passive, active) was randomized.

To investigate the effect of active arm movement on the perception of moving tactile stimuli, three arm movement conditions were introduced under which the tactile task was performed.

"Static" condition (Fig. 2.4): In this condition participants kept their arms in a rest position during tactile stimulation. A trial started by switching a rotary switch of the input device. After a delay between three and six seconds the sequential presentation of both pin-strokes (200 ms per pin-stroke) started. Both pin-strokes were separated by a 500 ms pause. Participants indicated whether the second stroke was shifted clockwise or counterclockwise with respect to the first stroke by switching the rotary switch of the input device, after which the next trial started automatically.

"Active" condition (Fig. 2.5): In the "active" condition participants performed an active arm movement. During the arm movement they received the tactile stimulation. From a resting position, which was adjacent to the subject, the setup had to be pushed forward within a velocity range of 0.2 and 0.3 m/s. This velocity had to be achieved within 10 cm from the starting position to trigger the pin movement. A velocity which was faster or slower than specified resulted in no tactile stimulus and the trial had to be redone. After entering their judgment using a custom made response box participants had to pull the arm of the force feedback device back to the starting position and the next trial was initiated.

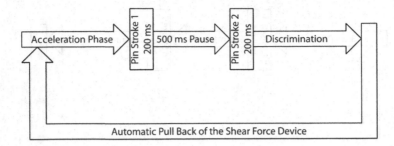

Fig. 2.6 Schematic flow chart of a trial in the "passive" condition

"Passive" condition (Fig. 2.6): The procedure in the "passive" condition was very similar to the "active" condition since the mechanical arm movements were identical. In the "passive" condition, however, the kinesthetic device applied a force resulting in a forward movement which drags the participant's arm along a trajectory. The force was increasing with the distance from the starting position to ensure a smooth acceleration. The goal was to achieve a velocity of 0.3 m/s within 8 cm from the starting position. The tactile strokes were executed after accelerating subject's arm to a velocity of 0.3 m/s and after covering the first 8 and 19 cm of the arm movement resulting in the same stroke output duration as in the other conditions. If subjects did not reach this velocity criterion within the given limits no stroke was elicited. This, however, never happened indicating that subjects did not actively slow down the robotic arm which means that there was no force applied by the user.

The participant's task was to discriminate the motion direction of two successively presented strokes on the finger pad. More precisely, they were instructed to report whether the second stroke was deviated clockwise or counterclockwise relative to the first stroke.

2.2.2 Results

The left panel of Fig. 2.7 depicts the mean thresholds of the three experimental conditions and the three standard directions. For a statistical analysis individual thresholds were entered into a 3×3 ANOVA with the factors "condition" (static, passive, active), and "direction" ($0°$, $45°$, $90°$). The main factor of "condition" was significant ($F(df, df) = 16.9$, $p < 0.001$). A subsequent t-test showed that there is a significant difference between the "static" and "passive" ($t(df) = -5.0$, $p < 0.001$) and between the "active" and "static" ($t(df) = 5.6$, $p < 0.001$) condition. The anisotropy for direction in any of the three conditions did not reach significance. That is, performance is equally worse within a given arm movement condition (active, passive, static), independent of whether the stroke direction is inline with the direction of arm movement or not. The right panel of Fig. 2.7 shows the same data collapsed over standard direction. Mean thresholds across subjects (for stroke direc-

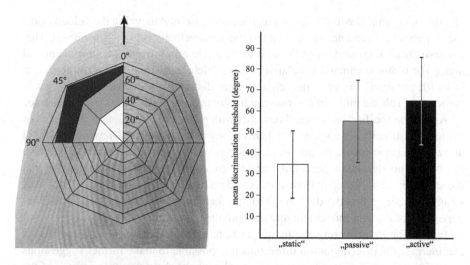

Fig. 2.7 *Left*: 84% discrimination thresholds in the coordinates of the finger tip (projected on the palm of the finger): Thresholds are depicted as a function of stroke direction. 0° corresponds to standard strokes in forward direction, 90° to standard strokes perpendicular to it. Thresholds are averages across participants. The *arrow* indicates the direction of the arm movement. *Right*: Mean across subject's discrimination thresholds and standard directions of each experimental condition. Error bars depict standard deviations of mean across subjects

tion) are 34.7° in the "static" condition, 55.6° in the "passive" condition and 65.4° in the "active" condition.

2.2.3 Discussion

The aim of this experiment was to investigate changes of tactile discrimination performance during arm movements. Our main finding is that tactile discrimination performance changes significantly due to active arm movement. This is in contrast to previous studies where discrimination thresholds were tested for electrical stimulation [24], vibrotactile stimulation [25] or tactile pattern discrimination [29]. In these studies, indications for an increase of discrimination thresholds were reported but they did not reach significance.

Our results show that arm movements have a considerable effect on tactile perception on the finger. Discrimination performance is best when the arm is unmoved and becomes worst when the arm is moved actively. An intermediate performance level is reached in the "passive" condition. Statistically, the discrimination performance in the "passive" condition does not differ from those in the "active" and "static" condition. This intermediate performance during "passive" motion is most likely caused by consciously or unconsciously moving the arm in order to follow the actively stirred robotic arm. An active arm movement, however, implies the presence of a motor command which causes a tactile suppression. Analyzing the velocity profile of passive movements shows that the specified velocity is always achieved. This suggests that the proportion of the unwanted active arm movement component and hence the motor command is relatively low which leads to a smaller suppression of tactile sensitivity. Even if participants were distinctively instructed not to act in favor of the robotic movements, one can hardly prevent participants from doing so.

Analyzing the discrimination thresholds with respect to the standard directions of stimulus motion shows no perceptual anisotropy at the fingertip. Drewing et al. [31] utilized the same tactile motion device as it is used in the actual study. They also discovered no significant perceptual anisotropy at the finger tip. Thus, the fact that the same device was used in both studies and no perceptual anisotropies were found strongly indicates that the discrimination performance at the finger tip is equal, independent of the arm movement and the stimulus direction.

In addition to answering the main question, namely, that active arm movement can increase the discrimination thresholds it is possible to make further suggestions regarding the origin of tactile suppression. The results show that the increase of the discrimination threshold is independent of the relative motion direction between the tactile stimulus and the arm. This supports the hypothesis that tactile suppression is the result of a gating effect, caused by a motor command, rather than the reafference principle. For the validation of the reafference principle it would have been expected to find a correlation of discrimination threshold change and arm movement direction. To get further evidence in favor of a gating mechanism to be responsible for tactile suppression, we need to show that tactile suppression depends on a collocation of tactile and kinesthetic information. From the actual results it is not possible to claim the involvement of a motor command in tactile suppression. The observed changes in the discrimination threshold could also be caused by the different workloads in the different arm movement conditions. In the "static" condition, participants had to perform the tactile task alone whereas in the active condition they also were instructed to move their arm within a specified velocity range. This can be seen as an additional task which may lead to a reduced performance in the tactile task during tactile arm movements. This issue will be investigated in Experiment 2 where the tactile task and the arm movement were performed contralaterally.

2.3 Experiment 2: Bilateral Control

Arm movements have been shown to affect the discrimination threshold. The results of Experiment 1 suggest an active arm motion (i.e., the involvement of muscle contraction) to be responsible for this decrease in tactile sensitivity. To strengthen this hypothesis, the possibility that an increased workload, associated with the arm movement, is responsible for the observed increase in discrimination threshold has to be excluded.

Two hypotheses will be tested in this experiment:

(1) The increase of the discrimination threshold in Experiment 1 may be due to a higher workload and participants are facing difficulties in a simultaneous conduction of both tasks (arm movement and tactile discrimination). If this were the case than by splitting the tasks between the different body sides should produce the same result as in Experiment 1. If anything this dual tasking should increase the workload and exaggerate the differences between the conditions.
(2) If, on the other hand, the observed discrimination threshold in Experiment 1 is caused by the motor command responsible for the arm movement, then the thresholds in all arm movement conditions should be the same, because the tactile task is now essentially independent from the motor task.

Here, all experimental conditions were identical to those in Experiment 1 except that the arm movement and tactile discrimination are separated to contralateral sides. The tactile task still has to be performed with the left index finger, while the right arm was moved.

2.3.1 Methods

Six right-handed, participants (three female) participated for pay. All of them were naïve to the purpose of the experiment. Their age ranged from 22 to 33 years (average 26 years). None of them reported previous injuries or impairment of tactile sensitivity of the finger tip.

Tactile discrimination performance was investigated by conducting the entire set of movement conditions ("active", "passive" and "still"—as described in Experiment 1) using the same kinesthetic and tactile display as in the previous experiment.

2.3.2 Results

The right panel of Fig. 2.8 shows the results for direction discrimination performance for the three arm movement conditions. Mean thresholds (for stroke direction) were 29.3° in the "static" condition, 35.9° in the "passive" condition and 35.1° in the "active" condition (Fig. 2.8). The mean discrimination thresholds were not statistically significantly different across the arm movement conditions.

The left panel of Fig. 2.8 depicts the mean thresholds of the three experimental conditions and the three standard directions. For statistical analysis individual thresholds were entered into a 3 × 3 ANOVA with the factors condition ("static", "passive", "active"), and Direction (0°, 45°, 90°). The main factor of condition was not significant ($F = 2.4$, $p = 0.158$). As in Experiment 1, there is no perceptual anisotropy for direction in any of the three conditions. That is, performance is equally worse with arm movement, independent of whether the stroke direction is in line with the direction of the arm movement or not.

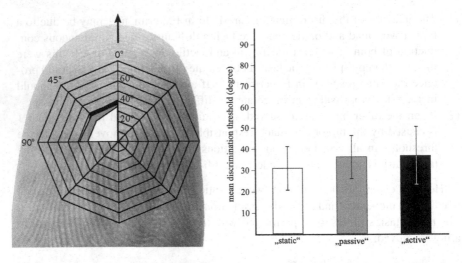

Fig. 2.8 *Left*: 84% discrimination thresholds in the coordinates of the finger tip (projected on the palm of the finger): Thresholds are depicted as a function of stroke direction. 0° corresponds to standard strokes in forward direction, 90° to standard strokes perpendicular to it. Thresholds are averages across participants. *Right*: Mean discrimination thresholds of each experimental condition averaged across standard direction. Error bars depict standard deviations across subjects and conditions

2.3.3 Discussion

This control experiment was conducted to provide further support of the results in Experiment 1, namely, that tactile suppression is mainly caused by a neural gating effect. The origin is supposed to be a motor command which is blocking the transition of tactile information from the peripheral neural system to the central neural system.

The results show that the tactile discrimination threshold at the finger tip does not depend on the direction of the tactile stimulus motion. In addition, discrimination thresholds do not show significant perceptual anisotropies with respect to the arm movement conditions. The mean discrimination thresholds across standard directions reveal that the discrimination performance is equally worse in all arm movement conditions. Here, the mean thresholds (ranging from 29.3° to 35.9°) are at the same level as the mean discrimination threshold of the "static" condition in Experiment 1 (34.7°). This strongly indicates that no tactile suppression occurred in the present experiment.

These results show that active arm movements do not induce tactile suppression per se. By comparing the results of the current and the previous experiments one factor can be identified that is supposed to be necessary to induce tactile suppression; this is the collocation of tactile and kinesthetic input. Only if both types of information are applied to the same body region, a significant drop in tactile sensitivity occurs. A displacement of the motor command to the contralateral side, as it is shown in the current experiment, results in the same discrimination threshold levels in either movement conditions.

In contrast to the previous experiment, the arm where the tactile stimulus was presented did not move. This means that there might have been a higher signal to noise ratio due to less neural firings during muscle and skin stretch, and there was no motor command which activates muscle contraction. Both are claimed to reduce tactile sensitivity [23, 32] by interfering with the tactile information and by inducing a gating effect, respectively. To identify which of those has a major effect on tactile sensitivity a follow up experiment has to be conducted where the time course of tactile suppression is investigated. The time course of both parameters, spontaneous firing rate and motor command, may show different characteristics. Sensory noise is mainly generated during the arm movement by muscle and skin stretch and, thus, tactile sensitivity should occur from the beginning of an arm movement. On the other hand, a motor command induced gating effect is supposed to occur before the arm movement since the motor command blocks neural pathways during its decent in order to activate the muscle and, thus, generates a block of afferent tactile information.

2.4 Experiment 3: Time Course of Tactile Suppression

In Experiment 1 and 2 it was shown that active arm movements decrease tactile discrimination performance. Furthermore, it was demonstrated that by splitting the tactile and kinesthetic task, discrimination thresholds in all arm movement conditions were essentially the same, suggesting a major impact of a motor command on the discrimination performance. However, to identify whether sensory noise or a motor command is generating tactile suppression, the time course of tactile suppression will be studied. In other studies, gating is reported to occur 60 ms to 100 ms before movement onset whereas others found gating only after movement onset [33–37]. Williams et al. [28] found a decrease in detection performance of weak electrical stimuli applied to a moving finger prior to its abduction which is reported to become significant at 120 ms before the movement onset.

Time courses have never been investigated with mechanical tactile stimuli. Here, a discrimination of pin position in combination with arm movements is investigated. By presenting the tactile stimuli before or during the arm movement it is expected to find changes in the discrimination threshold with respect to the presentation time. Two hypotheses will be examined:

(1) If a motor command is mainly involved in generating tactile suppression, then the gating effect will occur before the actual arm movement, because the motor command will be sent prior to the muscle contraction, resulting in a neural blockade of afferent tactile information. Thus, tactile suppression will be established before the motor command activates the muscle.
(2) If sensory noise is mainly responsible for tactile suppression, then a drop in tactile sensitivity will occur after the onset of an arm movement, because prior to the arm movement no additional sensory noise is expected.

2.4.1 Methods

Eight right-handed, participants (five male, three female) participated for pay. All of them were naïve to the purpose of the experiment. Their age ranged from 23 to 32 years (average 26 years). None of them reported previous injuries or impairment of tactile sensitivity of the finger tip.

In contrast to the previous experiments in this chapter, a different tactile display is used. The VirTouch Mouse (VTM) is a commercially available computer mouse containing three Braille modules (each consisting of a 4×8 pin matrix display) that extend plastic pins (diameter of 0.5 mm) normal to the skin surface. The range of pin movement is 1 mm, divided into 16 incremental steps. Figure 2.9 shows the VirTouch Mouse. Utilizing the VirTouch Mouse was necessary due to the need for a temporally short stimulus presentation to investigate the time course. A stimulus presentation consisted of an up and down movement of one pin (overall duration of 60 ms). In Experiment 1 and 2 the presentation of the tactile stimuli in a 2-interval forced-choice (2IFC) paradigm took 900 ms. In order to detect tactile sensitivity changes in a close temporal proximity to the onset of arm movements, this seems to be an inadequate method. Instead, a 1IFC paradigm is used where only one stimulus is presented per trial. This allows for sensitivity measurements in the temporal proximity of the arm movement onset.

The pin extension is regulated by bending piezo actuators. The interface, electrical drivers and the power supply, are integrated. To use the full functionality of the VirTouch Mouse as a tactile display, the development of a custom made device driver software was necessary. This device driver was realized as a RTLinux real time module under special privacy conditions from VTS (Virtual Touch Systems) regarding the transmission protocol of the serial link. In contrast to the previous experiments the VTM was mounted on a kinesthetic robot arm (DeKIFeD 4) in order to move and track participant's arm in space (for details see [38]). This kinesthetic device was built of the same electric motors and gears, resulting in similar kinematic properties as the kinesthetic device (ViSHaRD 10) earlier in this chapter. The free moving space of the kinesthetic device was restricted to one degree of freedom (1 DOF).

In this experiment, the "static", "active" and "passive" movement conditions are used. In the "active" condition participants had to perform an active arm movement during the pin discrimination task. After a brief tone, which was presented in a temporal window of three to six seconds after the trial start, participants pushed the setup, with a velocity of at least 0.2 m/s. As in the previous experiments, the "passive" arm movement condition was similar to the "active" condition. The crucial difference was that the kinesthetic device was producing a force in a forward or backward direction in order to move participant's arm along these directions. The "static" condition served as a baseline measurement, where no arm movement was required.

Participants were sitting on a bar stool in front of the kinesthetic/tactile setup. Their right hand was placed on the VirTouch Mouse and the distal phalanx of the left

Fig. 2.9 *Left*: Kinesthetic device (DeKIFeD 4) in combination with the tactile device (VirTouch Mouse). *Right*: Close-up view of the VirTouch Mouse containing three Braille modules and proportions of the Braille module. Distances are in millimeters

index finger was placed on the Braille module of the VirTouch Mouse. Two IBM-compatible PCs (one for the tactile and one for the kinesthetic device) controlled the stimulus presentation, data collection and the movement of the VirTouch Mouse in space, using custom programmed applications.

In the present experiment a discrimination task had to be performed where participants had to discriminate two spatially separated pins. The pins were adjacent and had a spatial distance of 1 mm (Fig. 2.9). Either the left or the right pin was lifted during each trial and participant's task was to name whether it was the left or the right one. Thus, the task is criterion free and a direct measure of sensitivity. The order of pin presentation and pin extension was randomized. Seven stimulus intensities were presented ranging from 5 to 11 on a scale of 12 possible steps (12 steps = 1 mm). The same custom made input device as described in the previous experiments was used to enter the responses. In the "active" and the "passive" arm movement condition, tactile stimulation was randomly presented in a temporal range of 200 ms before and 200 ms after the onset of the arm movement in steps of 1 ms. For the analysis the data was pooled in bins of 50 ms. The arm movement was performed for approximately one second.

2.4.2 Results

Figure 2.10 shows the mean proportion of correct answers across subjects. Answers were pooled in 50 ms bins. The dashed grey box depicts the time of the arm movement. The baseline performance of the current tactile discrimination task, which was measured in the "static" condition, was 82% of correct answers (Fig. 2.10, dotted

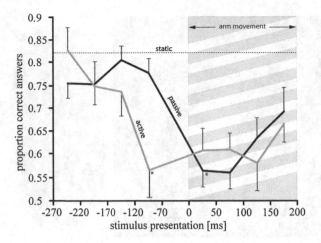

Fig. 2.10 Mean proportion of correct answers of tactile discrimination task in a "static" (*dashed line*), "active" (*grey line*) and "passive" (*black line*) condition. Stimulus presentation times are relative to the onset of arm movement (0 ms). The *dashed grey box* covers the time of the arm movement. Error bars depict standard errors. For demonstrational purposes only half error bars are displayed

line). Two main differences of the discrimination performance in the "active" and "passive" movement can be identified: First, the discrimination performance in the "active" condition (grey line) is continuously decreasing, beginning from the earliest stimulus presentation time (83% correct in the first bin). The lowest discrimination performance (57%) is observed in the time bin of 120 ms to 70 ms prior to the movement onset ($F(df, df) = 13.3$, $p < 0.01$). A subsequent t-test indicates a significant drop below the baseline from the time 120 ms to 70 ms prior to the arm movement ($t(df) = -4.3$, $p = 0.004$). After the movement onset, performance remains at a low level and even increases slightly, without reaching significance. Secondly, the "passive" condition (black line) shows a decrease in discrimination performance from a plateau (ranging from 75% to 81% correct answers), whereby the percentage of correct responses are not significantly different across the bins. The biggest drop (57% correct answers) can be observed in the time from the movement onset to 50 ms after movement onset ($F = 10.1$, $p = 0.016$). In this time bin the first significant drop occurs below the baseline ($t = -7.5$, $p < 0.001$). Subsequently, discrimination performance rises again but was not significantly different from the lowest performance level.

The proportion of correct answers reaches a maximum of 0.81 of the "passive" condition. A significant drop occurs in the time 0 ms/50 ms which corresponds to the onset of the passive arm movement. Here, the proportion of correct answers drops to a mean of 0.57. Tactile sensitivity represented by the proportion of correct answers increases after 100 ms but did not reach significance. By contrast, in the "active" condition, the proportion of correct answers drops significantly from an initial proportion of 0.83 to 0.57, occurring at a time of 70 ms to 120 ms prior to the arm movement. Also in this movement condition, tactile sensitivity rises after the arm movement has begun. Again, this rise does not reach significance.

2.4.3 Discussion

The goal of this experiment was to identify whether the decrease in tactile discrimination threshold can be addressed to a motor command that is generated during active arm movements or due to sensory noise caused by receptor activation by muscle and skin stretch during arm movement. Comparing the time course of tactile discrimination performance in the "active" and "passive" arm movement condition shows major differences: In the "passive" condition, a statistically significant drop in performance occurred from the beginning of the arm movement indicating that mainly sensory noise was interfering with the tactile target signal. A lower signal-to-noise ratio leads to a poorer discrimination performance. It may be speculated that a motor command interferes with the tactile information after the onset of arm movement. This will be further investigated in Experiment 4.

In the "active" arm movement condition, the strongest decrease in tactile discrimination performance appears in a time range between 70 ms to 120 ms prior to the arm movement. This speaks for a motor command to induce tactile gating and thereby establish a blockade at the neural border of the peripheral nervous system (PNS) and central nervous system (CNS). Such a blockade is supposed to decreases the amount of tactile information which is transmitted to the somatosensory cortex. Consequently tactile sensitivity is being decreased. Our results are in line with a neurophysiological study in awake monkeys by Seki et al. [23], who found that presynaptic inhibition is the result of descending motor commands, typically 400 ms before movement onset.

One observation in Fig. 2.10 provides an opportunity for speculation. In both arm movement conditions, the "active" and the "passive", a recovery of tactile discrimination performance occurs, starting between 50 ms and 150 ms after the onset of the arm movement. The performance increase may be explained by the end of the acceleration phase of the arm movement. A lower arm acceleration results in lower forces required from the arm and thus, a weaker motor command is involved.

In Experiment 1, tactile stimuli were presented about 300 ms after the onset of the arm movement, depending on the arm movement velocity. Nevertheless, a strong increase of tactile discrimination threshold can be observed. However, according to the results in Experiment 3, tactile sensitivity should be almost recovered by that time. Since the observed recovery of tactile discrimination did not show a statistically significant difference from the lowest performance near the onset of the arm movement this could be an artifact. Furthermore, a recovery of tactile sensitivity was never reported in other studies [28, 33–37, 39, 40].

In summary the results suggest that in the "active" arm movement condition a descending motor command causes a drop in tactile discrimination performance. In the "passive" arm movement condition a drop in tactile discrimination threshold can be observed only after the start of the arm movement. This drop may have two origins. First, sensory noise is originating in the muscles and the skin. Secondly, participants could support the forward movement of the kinesthetic device by actively moving their arms without taking notice. To distinguish between the two and pinpoint the origin of the drop in sensitivity we used electromyography (EMG) in a final experiment.

2.5 Experiment 4: Time Course of Muscle Contraction

It has been shown that active arm movements lead to a decrease in tactile sensitivity and that motor commands may play a major role in this. A decrease in tactile sensitivity during passive arm movement, which occurs after the onset of the movement, has to be further investigated to identify whether it originates from sensory noise or from unconscious muscle contraction. Here, an electromyographic (EMG) measurement is conducted to determine the timing of muscle activity and compare it to the drop in tactile sensitivity. The aim is to investigate whether the muscle activity precedes or follows the onset of the arm movement. Thus, two hypotheses will be investigated:

(1) A motor command which precedes the arm movement supports its suppressing effect on tactile performance.
(2) If the motor command does not proceed then it is unlikely that a motor command modulates tactile sensitivity and sensory noise.

2.5.1 Methods

Four right-handed, participants (three male) volunteered. All of them were naïve to the purpose of the experiment. Their age ranged from 25 to 38 years (average 30 years).

Subjects were sitting in front of the same setup as described in Experiment 3. They had to perform only the kinesthetic task. 40 trials were performed, half of them in the "active" arm movement condition and half of them in the "passive" arm movement condition. In the "active" condition subjects had to move the arm forward after a visual go signal had occurred. In the "passive" condition participants' arm was moved by the kinesthetic device.

EMG activity of the right Musculus triceps brachii (for two participants) or the right Musculus deltoideus (for the remaining two participants) was measured using surface electrodes of 10 mm in diameter (Fig. 2.11). Different subjects showed different muscle activities (M. triceps brachii vs. M. deltoideus). This could be due to differences in participant's posture which made it necessary to activate either muscle to move the arm forward. One pair of electrodes (12 cm center-to-center inter-electrodes distance) was placed centrally with respect to the muscles after cleaning

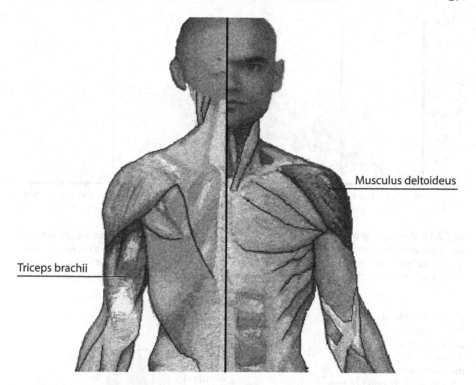

Musculus deltoideus

Triceps brachii

Fig. 2.11 Schematic diagram of muscles which electromyographic activity was measured. (Picture based on http://upload.wikimedia.org/wikipedia/commons/9/93/Deltoideus.png by Nikai, available under a Creative Commons Attribution-ShareAlike license)

the skin with alcohol, along a line parallel to the muscle fiber orientation. A ground electrode was attached to the subject's right upper arm. These electrodes were connected to an amplifier and afterwards via an AD-converter to a PC. The signal was band pass filtered from 3 Hz to 400 Hz by the amplifier.

A custom made MATLAB program (The MathWorks, Natick, Massachusetts, USA) was used to sample the potentials with a sample rate of 5 kHz in a time range of 300 ms before until 1000 ms after a trigger indicating muscle activity.

EMG data was pre-processed for each subject and each experimental condition (active and passive movement) separately. The AC component of the signal was rectified and then smoothed over 25 ms using the moving average. The baseline for thresholding was calculated for each trial independently: the data between the time point at which the velocity reached 1 cm/s and the trigger of the muscle activity was taken and the threshold for the trial was set to the mean of these data plus three standard deviations. The first time point at which the curve stayed over the threshold for at least 24 ms was taken as onset time of the response latency. Figure 2.12 shows a typical course of muscle activation in the "active" and "passive" condition. Muscle activation occurs 145 ms before the onset of arm movements in the active condition whereas in the passive condition it occurs 3 ms before (Fig. 2.13).

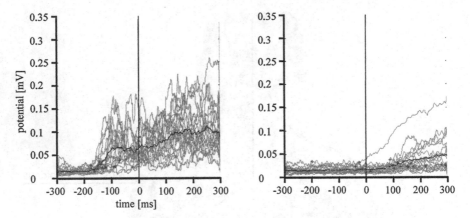

Fig. 2.12 An example of EMG recordings from the Triceps brachii of one subject for active (*left*) and passive (*right*) condition. *Light curves* (*red*) depict muscle activities of each trial. *Black curve* depicts the average activity potential across 20 trials. *Stars* (*blue*) depict the first supra-threshold activities. *Vertical black line* marks the onset of the arm movement

Fig. 2.13 EMG data depicts the average time of muscle activation onset across 4 subjects. Negative values on the *y*-axis indicate muscle activation before onset of arm movement. Error bars are standard errors

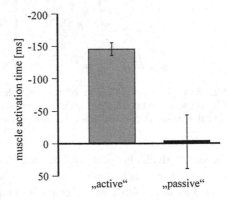

2.5.2 Results

2.5.3 Discussion

This experiment was conducted to define the origin of tactile suppression in the "passive" arm movement condition after the onset of arm movement. The results show that in the "active" condition there is steady muscle activation well before the onset of the actual arm movement. The mean activation time for both groups is 145 ms prior to the arm movement. In the "passive" condition, on the other hand, no muscle activity occurred prior to the arm movement. These results indicate that the decrease in tactile sensitivity in the "passive" condition after onset of an arm movement derives from sensory noise, since no muscle activation could be detected.

2.6 Conclusion

Tactile suppression is a phenomenon which describes a dampening of tactile sensitivity during active body movements. Today, the underlying mechanisms and the perceptual effects are still not satisfyingly understood. In this study it is shown that active arm movements clearly modulate tactile perception. An actively performed arm movement elicits a statistically significant drop in tactile discrimination performance compared to holding the arm still. Passive movements also lead to tactile performance decrease but to a lesser extent. Taken together, the results suggest an involvement of a descending motor command that is present during active arm movements but not in mere passive movement. Further proof for the involvement of a motor command in tactile suppression is given by a control experiment where tactile and kinesthetic task was split up to different body sides. No significant difference between the arm movement conditions was observed. Additionally, the investigation of the time course of tactile suppression shows that two major differences occur between the active and the passive arm movement condition. First, the discrimination threshold increases in the passive condition only as of the onset of the arm movement, suggesting that the sensory noise of the arm movement per se leads to the observed decrease in tactile sensitivity. Secondly, in the active condition an increase in tactile discrimination threshold is already measurable prior to the movement, which speaks for the involvement of a descending motor command that then causes a neural blockade that reduces tactile information transfer to the CNS.

Interpreting the results of this chapter supports the hypothesis of a descending motor command that blocks the neural afferent pathway and thereby limits the tactile information available to the CNS. Our results are in line with the results of [23] who found evidence for presynaptic inhibition via neurophysiologic experiments, which suppresses cutaneous input to the spinal cord during voluntary movements in primates. The two major sources of this inhibition are thought to come from descending central commands and peripheral inputs from afferent fibers, that is, sensory feedback from other cutaneous areas, muscle spindle and tendon organ afferents [41–44].

Suppression of tactile sensation may play an important functional role, namely filtering out task irrelevant tactile information during self-movement. Blakemore [17] claim that self-produced (actively) tactile stimuli are perceived less intensely than tactile stimuli that are generated by an unpredictable source, which agrees with our results. In other words one can say that tactile stimuli are less salient when they are accompanied by an active limb movement as when they are applied without self-movement. Blakemore, however, proposes predictions of the sensory consequences of self-induced movements to be compared to the actual sensory inputs. Subsequently, predictable sensory consequences will be cancelled out and only the differential signal of the prediction and the afferent signal will be processed.

The results of Blakemore suggest the involvement of the reafference principle in tactile suppression. Moreover, velocity, intensity and motion direction of tactile stimuli have to be predicted in order to explain Blakemore's results. The reafference principle, however, does not seem to have a major impact in our study, because the discrimination threshold on the finger tip is equally worse with respect of the standard direction of the stimulus motion. Here, stimulus motion which is orthogonal to the arm movements should be accentuated since during a forward arm movement only stimulus motions along the same axis should be expected. Hence, a perceptual anisotropy for the discrimination threshold would be expected to occur. But this did not happen, the reduction in sensitivity turned out to be not selective for the direction of stimulation. This difference between the two studies (Blakemore's and mine) may be explained by differences in the setup, the task and the experimental paradigm. That is, Blakemore used self- and externally generated tactile stimuli while the stimulated hand was unmoved. In our experiment the stimulus was always externally generated while the arm was moved actively, passively or was kept still.

We argue that tactile suppression during active exploration of, for example surfaces, is not desirable since (relevant) tactile features would be blurred out. However, tactile suppression may reduce uninformative tactile information from task irrelevant skin stretch and muscle contraction. Filtering out these task irrelevant information can help to optimize the signal to noise ratio and, thus, to highlight tactile features of an actively explored object.

Experiment 2 was conducted to control whether tactile suppression could be explained by an increased workload during arm movement compared to the "static" condition. By separating the tactile and motor tasks between the left and right hand, respectively, we were able to corroborate the results in Experiment 1. The results rule out an explanation in terms of changes in workload since the workload was similar to that in Experiment 1. The results show only a small but non-significant increase of the discrimination threshold in the "passive" and "active" condition compared to the "static" condition. A modest decrease in tactile sensitivity by splitting the tactile and motor task is also reported by [28] and supports our finding even if the decrease does not reach significance.

Experiment 3 provides insight in the time course of tactile suppression and identifies the main components of tactile suppression. Both the active and passive movement conditions show a significant decrease of tactile sensitivity around the time of the arm movement onset (Fig. 2.10). The main difference between these movement conditions is the beginning of the sensitivity drop. In the "active" condition the sensitivity drops 70 to 120 ms prior to the arm movement. In the "passive" condition, on the other hand, it does not start until the onset of the actual arm movement. The timing of this gating effect varies between studies. It was reported to occur 60 to 100 ms before movement onset [33, 34] whereas others found gating only after movement onset [35–37]. In agreement with our results Williams et al. [28] found a decrease in detection performance of weak electrical stimuli applied to a moving finger prior to its abduction which is reported to become significant at 120 ms before the movement onset.

The inconsistencies between the various studies can be partly explained by differences in the setup and protocol design, including such factors as the type of stimulus that was used, whether near-threshold or supra-threshold stimuli were applied, whether magnitude estimation or detection tasks were conducted, and whether limb movement was purposeful or spontaneous. Seki et al. [23] gained evidence that tactile afferent input to spinal cord interneurons is blocked during active wrist movement in awake monkeys. They claim that a presynaptic inhibition is effectively produced by descending motor commands. Their findings are congruent with our results showing that a descending motor command, which is present only in the "active" condition, is supposed to be responsible for a significant drop in tactile discrimination performance prior to the movement onset (Fig. 2.10).

It is remarkable that tactile sensitivity seems to recover some time after the onset of the arm movement. The increase is evident in both arm movement conditions although it does not reach significance and assumptions about the origin are speculative. According to Williams et al.'s [39] model of detection threshold over time, tactile sensitivity is supposed to decrease after the movement onset. Recovering is not supposed to happen. The rate at which sensitivity is expected to decrease depends on the intensity of the tactile stimulus. The higher the intensity of tactile stimulation is, the longer it takes to lower the sensitivity. However, interpreting the observed sensitivity increase, one can speculate on the behavioral relevance. During tactile exploration, for example, one intuitively assumes that tactile sensitivity is supposed to be as high as possible in order to detect as many tactile features as possible. Tactile suppression, however, antagonizes an exploration task by gating out tactile stimuli with low intensities. Presumably, re-sensitizing occurs after the movement onset, that is, after the transition from rest to movement, which might be relevant for masking unwanted sensory signals but provide a lower discrimination threshold during movement. However, the results of Experiment 1 do not support this speculation. There, tactile stimuli were applied 300 ms after the onset of arm movements and a high level of tactile suppression was measured. According to the time course of Experiment 3, an almost full recovery of tactile sensitivity would be expected by then. This contradiction speaks for a random effect which is observed in the time course in Experiment 3.

Furthermore, the drop in tactile sensitivity after the onset of an arm movement can be attributed to sensory noise masking the tactile target stimulus. This assumption is supported by previous studies [42–44] claiming that tactile performance becomes worse *during* movement because of sensory feedback evoked by skin and muscle deformation. Thus, peripheral sensory feedback increases which creates a drop of tactile sensitivity.

The final control experiment allowed for the observation that EMG activity precedes the arm movement by the same time as a drop in tactile discrimination performance was observed to drop. Previous studies have shown that a suppression of afferent information during active movement precedes EMG onset, providing a hint for the dominant role of descending motor commands in generating presynaptic inhibition of afferent input [23]. Unfortunately, the participation of different subjects in our experiments did not allow us to correlate the time of EMG onset and the drop in discrimination performance.

In conclusion, the results strongly suggest the involvement of motor commands, generated during voluntary movements, in tactile suppression. This main conclusion is supported by the fact that voluntary movements on different limbs do not affect the tactile discrimination performance. Thus, this is the first evidence that active movement does not only affect the detection threshold but also the discrimination threshold. Another finding that supports the involvement of a motor command in tactile suppression is that tactile sensitivity decreases before the arm movement and that the electromyografic signal was detected before the muscle contraction. This indicates the presence of the motor command prior to the muscle activity which is known to induce a gating effect of the afferent pathway leading to reduced tactile information in the CNS and, thus, to a poorer representation of tactile stimuli and to poorer discrimination abilities.

Acknowledgements We thank Martin Buss and Angelika Peer for providing us with the ViSHaRD10 and the DeKIFeD4 devices and the help for setting up these systems in our lab. Furthermore, we thank Michael Fritschi and Mario Kleiner with help in the programming of the experimental procedure, Massimiliano Di Luca for help in conceptualizing the experiment and Alexandra Reichenbach and Axel Tielscher for help with the EMG analysis. This work was partly supported by the ImmerSence project within the 6th Framework Programme of the European Union, FET—Presence Initiative, contract number IST-2006-027141, see also www.immersence.info.

References

1. Kastner, S., De Weerd, P., Desimone, R., Ungerleider, L.G.: Mechanisms of directed attention in the human extrastriate cortex as revealed by functional mri. Science **282**(5386), 108–111 (1998)
2. Burr, D.C., Holt, J., Johnstone, J.R., Ross, J.: Selective depression of motion sensitivity during saccades. J. Physiol. **333**, 1–15 (1982)
3. Burr, D.C., Morrone, M.C., Ross, J.: Selective suppression of the magnocellular visual pathway during saccadic eye movements. Nature **371**(6497), 511–513 (1994)
4. Richards, W.A.: Saccadic suppression. J. Opt. Soc. Am. **59**, 617–623 (1969)
5. Helmholtz, H.: Handbuch der physiologischen Optik. Voss, Leipzig (1867)
6. von Holst, E., Mittelstaedt, H.: Das Reafferenzprinzip. Naturwissenschaften **20**, 464–475 (1950)
7. Ilg, U.J., Hoffmann, K.P.: Motion perception during saccades. Vis. Res. **33**(2), 211–220 (1993)
8. Matin, E.: Saccadic suppression: a review and an analysis. Psychol. Bull. **81**(12), 899–917 (1974)
9. Sperry, R.W.: Neural basis of the spontaneous optokinetic response produced by visual inversion. J. Comp. Physiol. Psychol. **43**(6), 482–489 (1950)
10. Holt, E.: Eye movement and central anesthesia. Harv. Psychol. Stud. **1**, 3–45 (1903)
11. Mackay, D.M.: Elevation of visual threshold by displacement of retinal image. Nature **225**(5227), 90–92 (1970)
12. Burr, D.C., Morgan, M.J., Morrone, M.C.: Saccadic suppression precedes visual motion analysis. Curr. Biol. **9**(20), 1207–1209 (1999)
13. Bridgeman, B., Hendry, D., Stark, L.: Failure to detect displacement of the visual world during saccadic eye movements. Vis. Res. **15**(6), 719–722 (1975)
14. Shioiri, S., Cavanagh, P.: Saccadic suppression of low-level motion. Vis. Res. **29**(8), 915–928 (1989)
15. Diamond, M.R., Ross, J., Morrone, M.C.: Extraretinal control of saccadic suppression. J. Neurosci. **20**(9), 3449–3455 (2000)
16. Blakemore, S.J., Frith, C.D., Wolpert, D.M.: Spatio-temporal prediction modulates the perception of self-produced stimuli. J. Cogn. Neurosci. **11**(5), 551–559 (1999)
17. Blakemore, S.J., Goodbody, S.J., Wolpert, D.M.: Predicting the consequences of our own actions: the role of sensorimotor context estimation. J. Neurosci. **18**(18), 7511–7518 (1998)

18. Blakemore, S.J., Wolpert, D., Frith, C.: Why can't you tickle yourself? NeuroReport **11**(11), 11–16 (2000)
19. Blakemore, S.J., Wolpert, D.M., Frith, C.D.: Central cancellation of self-produced tickle sensation. Nat. Neurosci. **1**(7), 635–640 (1998)
20. Blakemore, S.J., Wolpert, D.M., Frith, C.D.: The cerebellum contributes to somatosensory cortical activity during self-produced tactile stimulation. NeuroImage **10**(4), 448–459 (1999)
21. Rushton, D.N., Rothwell, J.C., Craggs, M.D.: Gating of somatosensory evoked potentials during different kinds of movement in man. Brain **104**(3), 465–491 (1981)
22. Chapman, C.E., Beauchamp, E.: Differential controls over tactile detection in humans by motor commands and peripheral reafference. J. Neurophysiol. **96**(3), 1664–1675 (2006)
23. Seki, K., Perlmutter, S.I., Fetz, E.E.: Sensory input to primate spinal cord is presynaptically inhibited during voluntary movement. Nat. Neurosci. **6**(12), 1309–1316 (2003)
24. Chapman, C.E., Bushnell, M.C., Miron, D., Duncan, G.H., Lund, J.P.: Sensory perception during movement in man. Exp. Brain Res. **68**(3), 516–524 (1987)
25. Post, L.J., Zompa, I.C., Chapman, C.E.: Perception of vibrotactile stimuli during motor activity in human subjects. Exp. Brain Res. **100**(1), 107–120 (1994)
26. Schmidt, R.F., Schady, W.J., Torebjork, H.E.: Gating of tactile input from the hand. I. Effects of finger movement. Exp. Brain Res. **79**(1), 97–102 (1990)
27. Schmidt, R.F., Torebjork, H.E., Schady, W.J.: Gating of tactile input from the hand. II. Effects of remote movements and anaesthesia. Exp. Brain Res. **79**(1), 103–108 (1990)
28. Williams, S.R., Shenasa, J., Chapman, C.E.: Time course and magnitude of movement-related gating of tactile detection in humans. I. Importance of stimulus location. J. Neurophysiol. **79**(2), 947–963 (1998)
29. Lamb, G.D.: Tactile discrimination of textured surfaces: psychophysical performance measurements in humans. J. Physiol. **338**, 551–565 (1983)
30. Ueberle, M., Mock, N., Buss, M.: Vishard10, a novel hyper-redundant haptic interface. In: 12th International Symposium on Haptic Interfaces for Virtual Environment and Teleoperator Systems (HAPTICS'04) (2004)
31. Drewing, K., Fritschi, M., Zopf, R., Ernst, M.O., Buss, M.: Tactile display exerting shear force via lateral displacement. ACM Trans. Appl. Percept. **2**(2), 118–131 (2005)
32. Faisal, A.A., Selen, L.P., Wolpert, D.M.: Noise in the nervous system. Nat. Rev., Neurosci. **9**, 292–303 (2008)
33. Chapman, C.E., Jiang, W., Lamarre, Y.: Modulation of lemniscal input during conditioned arm movements in the monkey. Exp. Brain Res. **72**(2), 316–334 (1988)
34. Cohen, L.G., Starr, A.: Localization, timing and specificity of gating of somatosensory evoked potentials during active movement in man. Brain **110**(Pt 2), 451–467 (1987)
35. Dyhre-Poulsen, P.: Perception of tactile stimuli before ballistic and during the following manner tracking movements. In: Gordon, G. (ed.) Active Touch, pp. 171–176. Pergamon, Oxford (1978)
36. Ghez, C., Pisa, M.: Inhibition of afferent transmission in cuneate nucleus during voluntary movement in the cat. Brain Res. **40**(1), 145–155 (1972)
37. Jiang, W., Lamarre, Y., Chapman, C.E.: Modulation of cutaneous cortical evoked potentials during isometric and isotonic contractions in the monkey. Brain Res. **536**(1–2), 69–78 (1990)
38. Kron, A., Schmidt, G.: Haptic telepresent control technology applied to disposal of explosive ordnances: Principles and experimental results. In: Proceedings of the IEEE International Symposium on Industrial Electronics (ISIE), pp. 1505–1510 (2005)
39. Williams, S.R., Chapman, C.E.: Time course and magnitude of movement-related gating of tactile detection in humans. II. Effects of stimulus intensity. J. Neurophysiol. **84**(2), 863–875 (2000)
40. Williams, S.R., Chapman, C.E.: Time course and magnitude of movement-related gating of tactile detection in humans. III. Effect of motor tasks. J. Neurophysiol. **88**(4), 1968–1979 (2002)
41. Eccles, J.C., Schmidt, R.F., Willis, W.D.: Depolarization of the central terminals of cutaneous afferent fibers. J. Neurophysiol. **26**(4), 646–661 (1963)

42. Janig, W., Schmidt, R.F., Zimmermann, M.: Two specific feedback pathways to the central afferent terminals of phasic and tonic mechanoreceptors. Exp. Brain Res. **6**(2), 116–129 (1968)
43. Jankowska, E., Slawinska, U., Hammar, I.: Differential presynaptic inhibition of actions of group II afferents in di- and polysynaptic pathways to feline motoneurones. J. Physiol. **542**(Pt 1), 287–299 (2002)
44. Rudomin, P., Schmidt, R.F.: Presynaptic inhibition in the vertebrate spinal cord revisited. Exp. Brain Res. **129**(1), 1–37 (1999)

Chapter 3
Weight Perception with Real and Virtual Weights Using Unimanual and Bimanual Precision Grip

Christos D. Giachritsis and Alan M. Wing

Abstract Accurate weight perception is crucial for effective object manipulation in the real world. The design of force feedback haptic interfaces for precise manipulation of virtual objects should take into account how weight simulation is implemented when manipulation involves one hand or two hands with one or two objects. The importance of this is apparent in tasks where the user requires applying vertical forces to penetrate a surface with a tool or splice a fragile object using one and/or two hands. Accurate perception of simulated weight should allow the user to execute the task with high precision. Nonetheless, most commonly used force feedback interfaces for direct manipulation of virtual object use a thimble through which the users interacts with the virtual object. While this allows use of proprioceptive weight information, it may reduce the cutaneous feedback that also provides weight information through pressure and skin deformation when interacting with real objects. In this chapter, we present research in unimanual and bimanual weight perception with real and virtual objects and examine the magnitude of grip forces involved and relative contribution of cutaneous and proprioceptive information to weight perception.

3.1 Introduction

Effective object manipulation requires accurate weight perception since weight information will allow us to decide about a number of handling parameters including grip force, unimanual or bimanual mode and body posture. For example, a heavy object may require a stronger grip, a supporting hand posture and/or the use of both hands to lift, explore it or displace it. Moreover, accurate weight perception is important in tasks where precise application of forces to penetrate a surface with a tool

C.D. Giachritsis (✉)
Research and Development, BMT Group Ltd., Goodrich House, 1 Waldegrave Road, Teddington, London TW11 8LZ, UK
e-mail: cgiachritsis@bmtmail.com

A.M. Wing
School of Psychology, University of Birmingham, Edgbaston, Birmingham B15 2TT, UK
e-mail: a.m.wing@bham.ac.uk

A. Peer, C.D. Giachritsis (eds.), *Immersive Multimodal Interactive Presence*,
Springer Series on Touch and Haptic Systems,
DOI 10.1007/978-1-4471-2754-3_3, © Springer-Verlag London Limited 2012

(e.g., surgical scalpel) or assemble fragile objects is required. Research has shown that weight perception is based on *cutaneous* and *proprioception* signals [1, 2]. Proprioception signals originate primarily from the muscular activity required to displace (lift, move and land) an object, while cutaneous signals originate from the deformation of the skin resulting from grasping and holding the object. Mc-Closkey [1] found that when participants held weights with anaesthetised fingers (thumb, index and middle) and had their arms rested (i.e., no proprioceptive and no cutaneous cues), they were unable to make reliable weight judgments; that is, they could not match weights presented to the anaesthetised hand to the control weights lifted by the indicator hand which maintain both proprioception and cutaneous cues. Moreover, the contribution of cutaneous signals has been demonstrated indirectly by studies in which the weight percept was affected by the roughness of the object surface [3, 4], its size [1, 5] and the number of digits involved in grasping [5]. For example, Flanagan et al. [3] found that weight perception was affected by the surface texture of the lifted object: the smoother the surface the heavier the perceived weight of the object. This effect was attributed to the greater grip forces applied to objects with smoother surfaces [4] in order to secure an effective grip. In addition to the surface texture, the size of the object seems to affect weight perception. For example, McCloskey [1] and Flanagan and Bandomir [5] found that the bigger the object the lighter its perceived weight. This effect has been attributed to the classic size-weight illusion [6, 7] and has been found to be independent of grip force [8]. The contribution of proprioceptive signals has been demonstrated by studies in which 'sense of effort' and weight perception were affected by fatigue, changes in muscle condition, paresis, arm weight and posture [6–9]. Gandevia and McCloskey [6] found that anaesthesia and electrical stimulation of a group of muscles can result in different weight percepts. For example, when lifting a weight by flexing the index finger, participants reported that the weight felt heavier when the thumb was anaesthetised and lighter when the thumb was electrically stimulated. Weights also feel heavier when neuromuscular transmission is impaired as in the case of paretic patients [7]. Moreover, Ross, Brodie and Benson [8] found that in weight-lessness sensitivity to weight changes reduces. In another experiment, Wing, Giachritsis and Roberts [9] found that when participants held weights with their arm fully extended to the side forming a 90° angle with their torso (e.g., maximum torque), they judge them to be heavier than similar weights held with their arm at relaxed position forming a 0° angle with their torso (e.g., no torques).

Most current non-wearable desktop haptic interfaces, which are designed for object manipulation, employ force feedback to simulate shape and weight (e.g., PHANToM, Delta, Omega, etc.). Moreover, object exploitation and manipulation is based on single-point contact design either through a probe or thimble. This hardware design limits the realism of object manipulation and weight simulation since in real life we manipulate objects by grasping them which requires more than one fingers and our fingers are in direct contact with the object. As a result, object exploration can be performed only sequentially while no information about shape and/or surface texture can be acquired through the skin. A number of solutions have been suggested to overcome the limitation of the single-point contact interface either by combining more than one single-point contact devices, such as PHANToM 1.5, or designing new haptic interfaces, such as MasterFinger2 (MF2) and HIRO-II [10–14]. Combining more than one device can result in loss of working space, a drawback avoided by new interfaces such as MasterFinger2 and HIRO-II. Moreover, it is possible combining more than one purpose-built multi-point contact interfaces to enable bimanual manipulation of virtual objects. Nonetheless, even though the number of contact points increase, the user still grasps and manipulates virtual objects using thimbles. Therefore, the problem of lack of cutaneous feedback still remains. In this chapter, we report studies which investigated the effectiveness of MF2 in simulating unimanual and bimanual weight perception. We compare the results with those of studies involved real weights and we discuss the grip forces involved and the relative contribution of cutaneous and proprioception feedback.

3.1.1 Experiment 1: Evaluation of Unimanual and Bimanual Weight Perception During Manipulation of Virtual Objects

In this section we report a study which aimed to evaluate weight perception during unimanual and bimanual manipulation of virtual objects simulated with the new desktop haptic interface Master Finger 2 (MF2) [13]. The experiment was designed to collect data on unimanual and bimanual weight perception and then compare them with data from a similar experiment which used real weights [15]. The data have been previously reported in [16].

Methods

Apparatus and Stimulus The MF2 device was used to generate virtual weights. The users could manipulate the virtual objects using unimanual and bimanual precision grip. Also, they had visual feedback of their interactions in the desktop virtual environment. The visual feedback included both a virtual representation of their hands and box (Fig. 3.1).

A constant stimuli procedure [17] was employed to obtain weight discrimination thresholds for virtual weights. The MF2 haptic interface was used to generate a virtual box with dimensions 100 mm × 170 mm × 150 mm (WxDxH). The box width determined the width of the grip while the box depth determined the maximum distance between left and right grip in the bimanual lifting. The advantage of short depth is that it minimises the risk of torques affecting weight judgements in the unimanual lifting by allowing the user to grasp the box near the virtual centre of mass. The minimum reliable virtual weight was 75 g and was used as the minimum step size (i.e., the minimum difference two virtual weights that the users would have to perceive). In total, seven virtual weights were tested: 75, 150, 225, 300, 375, 450 and 525 g. The mid-range weight of 300 g was used as the standard weight. The users would have to compare the standard weight against all weights of the range and provide a judgement about their difference (i.e., light or heavier).

Fig. 3.1 The MF2 desktop virtual environment in which the users performed unimanual and bi-manual object lifting. The physical position of their hands relatively to the virtual weights was represented by the position of virtual hands relatively to the position of the virtual box

3.1.2 Procedure

Eight right-handed postgraduate students (average age of 26) of the Universidad Politécnica de Madrid were volunteered to participate in the experiment. They placed their index and thumb of both hands into the MF2 thimbles to be able to use their precision grip for manipulation. Furthermore, they guided their grasp through visual feedback of their hands and the virtual box (Fig. 3.1). A temporal *two-interval-forced-choice* (2IFC) paradigm with two conditions similar to the one used with real weights in [15] was also used here. The two conditions were: the experimental condition which tested weight discrimination between the right hand and both hands (BH) and the control condition which tested weight discrimination within the right hand (RH). Each trial consisted of three phases: in the first phase, users lifted the test/standard weight with their right hand/both hands (in the experimental condition) and right hand (in the control condition); in the second phase, they lifted the standard/test weight with their right hand/both hands (in the experimental condition) or with their right hand (in the control condition); and, in the third phase, they verbally reported which weight felt heavier, first or the second? (see Table 3.1).

The presentation sequence of standard and test weights was balanced and the trials were randomised for each session. Each test weight was compared twelve times against the standard weight. Therefore, each participant responded 168 times in the experimental condition and 84 times in the control condition. Each session lasted approximately 45 minutes and included a short break midway.

Table 3.1 The three phases of the 2IFC procedure employed (STD = standard weight; TEST = test weight)

Condition	Phase 1	Phase 2	Phase 3
Experimental	RH lift STD/TEST or BH lift STD/TEST	BH lift TEST/STD or RH lift TEST/STD	Which weight was heavier? Phase 1 or Phase 2?
Control	RH lift STD/TEST	RH lift TEST/STD	

Results

The evaluation of performance with real and virtual data was carried out by applying a psychometric function on individual data from both experimental and control conditions. The commonly used psychometric function $\Psi(x)$,

$$\Psi(x, \alpha, \beta, \gamma, \lambda) = \gamma + (1 - \gamma - \lambda)F(x, \alpha, \beta) \tag{3.1}$$

where γ is the lower and $1 - \gamma$ is the upper bound of the function [18]. $F(x, \alpha, \beta)$ is the two-parametric logistic function

$$F(x, \alpha, \beta) = \frac{1}{1 + e^{-(x-\alpha)/\beta}} \tag{3.2}$$

where α is location of the function on the x-axis and β is its slope. All individual and overall data were fitted with this psychometric function. The location of the function, α, relates to the point of *subjective equality* (PSE) while the slope, β, related to the *discrimination threshold* (DT) [17]. The PSE and DT for both the experimental and control functions were calculated and compared. The DT was taken to be half the absolute difference between the 25% and 75% thresholds (i.e., $|T25 + T75|/2$). Figure 3.2 illustrates an example of fitting the psychometric function and obtaining PSE and DT from the data of user U3. The inverse of the control function for the RH(test)-BH(std) stimulus presentation was used to compare data against the RH(std)-BH(test) stimulus presentation.

Results showed a double effect of bimanual manipulation on weight perception similar to the one observed in [15] (Fig. 3.3). First, virtual weights lifted with the right hand felt heavier than virtual weights lifted with both hands. A paired T-test showed that this effect was statistically significant with real weights when the standard weight was lifted with either the right hand ($T_{(5)} = 7.595$, $p = 0.001$) or both hands ($T_{(5)} = -4.233$, $p = 0.008$). However, with virtual weights this effect was found to be statistically significant only when the standard weight was lifted with the right hand ($T_{(7)} = 2.821$, $p = 0.026$). Second, sensitivity to changes in virtual weight reduced in the experimental condition. A paired T-test showed that this effect was statistically significant with real weights when the standard weight was presented in the right hand ($T_{(5)} = -4.321$, $p = 0.008$) and both hands ($T_{(5)} = 2.62$, $p = 0.047$). Similarly, the effect on DT was statistically significant with virtual

Fig. 3.2 Example of fitting
the psychometric function on
data from user U3. Each data
point is based on 12
responses. The overall PSE
and DT for the experimental
(exp) and control (ctrl)
conditions were obtained and
compared

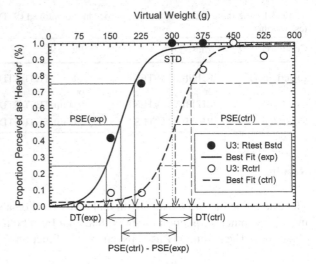

weights when the standard weight was presented in the right hand ($T_{(7)} = -4.572$, $p = 0.003$) and both hands ($T_{(7)} = 2.732$, $p = 0.029$).

Moreover, a comparison of the real and virtual weight control conditions (Fig. 3.4) showed that user's discrimination of virtual weights is not as accurate as with real weights. The DT with real weight is about 8 g while the DT with virtual weights is about 56 g. Even though this is lower than the step size of 75 g that was used in the experiment it is still nearly six times higher than the DT of about 9.8 g predicted by Weber's Fraction.

Discussion

The experiment described here showed that the discriminating between virtual weights lifted with one and two hands resulted in the same effect observed with real weights. That is, the bimanually lifted virtual weights tended to feel lighter than the unimanually lifted weights. However, the effect was not as prominent as the effect which was observed previously using real weights [15]. In addition, Weber's Fraction obtained from the right hand psychometric functions with both real and virtual weights showed that sensitivity to weight discrimination was reduced with virtual weights: users seem to be five times less sensitive to virtual than real weights. These results may be explained on the basis of both cutaneous and proprioceptive cues. For example, the users had no reliable cutaneous feedback about the weights and therefore relied exclusively on proprioceptive information to estimate the virtual weight. Therefore, the lack of cutaneous information could have resulted in a deterioration of weight sensitivity. Furthermore, users may be sensitive to small adjustments of horizontal forces exerted by the MF2 which are necessary to maintain a constant virtual size of the box during lifting. These additional proprioceptive signals, which were irrelevant to the task, could have interfered with proprioceptive

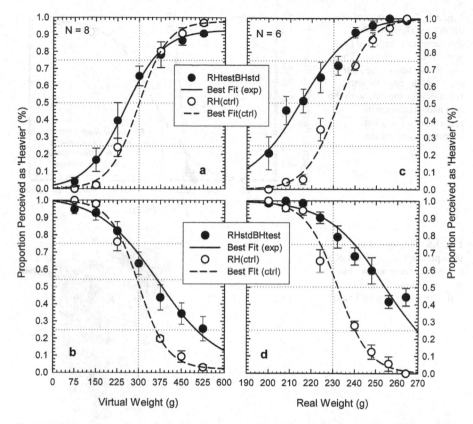

Fig. 3.3 Overall data and psychometric functions obtained for the experimental (● with *continuous lines*) and control (○ with *dashed lines*) conditions under both presentation orders of the virtual (**a, b**) and real weights (**c, d**) from [15]. Error bars represent standard error

information from vertical forces simulating weight. Similarly, during bimanual lifting there may be a discrepancy between the forces delivered to both hands by the MF2 and the forces expected by the user based on individual hand position and lifting effort. Nonetheless, the present MF2 set up has managed to simulate effectively unimanual and bimanual weight sensation and establish a clear presence for both.

3.1.3 Experiment 2: Bilateral Weight Perception with Virtual Weights

In experiment 1, it was shown that virtual weights lifted with both hands felt 'lighter' than similar virtual weights lifted with one hand. This result was consistent with studies which used real weights [15]. There could be at least two possible explanations for this effect. First, increasing the number of digits could decrease the total

Fig. 3.4 Comparison of unimanual (RH) weight discrimination with real (•) and virtual (o) weights. The *continuous* and *dashed curves* represent the psychometric functions for the data with the real and virtual weights, respectively. Error bars represent standard error

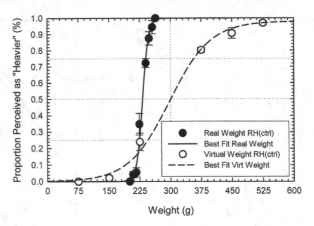

Fig. 3.5 The apparatus used in experiment 2 consisted of a force transducer attached to the top of a box in which different weights were loaded

grip force required to lift and hold the object [19, 20] resulting in a 'lighter' percept [5]. Second, sharing the weight between the two hands could also result in a 'lighter' percept given that some contralateraly accessible sensorimotor weight information could be lost due to lateralisation [21]. Here, we carried out a study to investigate whether the bimanual 'lighter' weight judgements are based on a reduced bimanual grip force due to doubling the digits involved in the grip. The study has been previously reported in [22].

Methods

Apparatus and Stimulus   Three weights were loaded on a box attached to a Novatech 200N force transducer which measured forces normal to the surface where the grip was applied (Fig. 3.5). The total weight of the apparatus was 200, 232 and 264 g. The width of the grip was 55 mm and its sides were covered with a fine (320 grade) sand paper to provide a firm and comfortable grip. The grip forces were recorded using National Instruments LabVIEW software.

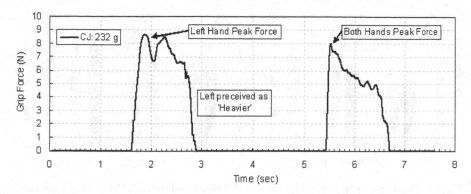

Fig. 3.6 Example of recorded grip forces when lifting the same weight (232 g) with the left hand and then with both hands (participant CJ)

Procedure   Six undergraduates students from the University of Birmingham, aged between 20 and 22, volunteered to take part in the study. The students were paid for their participation. They were instructed to use their precision grip to lift a single weight (200, 232, or 264 g) consequently with one hand (RH/LH) and both hands (BH) and then they reported which weight felt heavier. This procedure was similar to the procedure of experiment 1 but only tested three rather than seven weights. Therefore, no psychometric function was obtained. Before the start of the session, they had been informed that they would be lifting two different weights. The weight was lifted for about 100 mm and then was returned to its resting position immediately. Each weight was tested 12 times, the presentation order was balanced and the trials were randomised.

Results

The unimanual (RH/LH) and bimanual grip forces were recorded and their peak forces were averaged across weights and compared. Figure 3.6 shows an example of recorded grip forces and their peaks during a single trial.

Figure 3.7 shows the average grip forces for the LH-BH and RH-BH presentations. It is apparent that there was no indication that total unimanual grip force was greater than the bimanual grip force. The RH grip force appeared to be greater, than the BH but a T-test showed no statistical significance.

Moreover, the proportion of trials in which the unimanually (LH/RH) and the bimanually (BH) lifted weights were perceived as 'heavier' were averaged and compared. Figure 3.8 shows that weights lifted with the LH or RH felt heavier than weights lifted with BH. That was statistically significant (for $\alpha = 0.05$) with all three weights in LH condition and with the 232 g weight in RH condition.

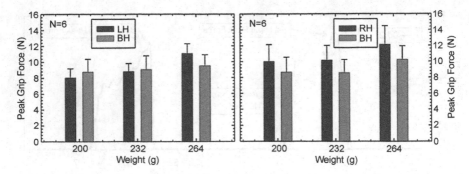

Fig. 3.7 Overall peak grip force during unimanual (LH/RH) and bimanual (BH) lifting. Error bars represent standard error

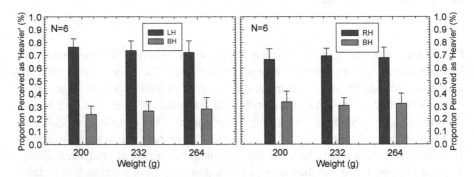

Fig. 3.8 Proportion of weights perceived as 'heavier' during unimanual (LH/RH) and bimanual (BH) lifting. Error bars represent standard error

Discussion

The results showed mainly that weights lifted with one hand felt heavier than weights lifted with both hands which are consistent with the results reported in experiment 1 and [15]. However, the total peak bimanual grip force was not significantly different than the unimanual peak grip force. This suggests that the 'lighter percept' may be the product of an imperfect integration of centrally stored weight information from both hands, which may be due to lateralisation [21], rather than the product of variable peripheral information signalling reduced effort due to a more effective bimanual grip.

3.1.4 Experiment 3: Relative Contribution of Cutaneous and Proprioceptive Information

In experiment 1, we saw that bimanual lifting results in 'lighter' weight estimates and that this effect could be effectively simulated by the MF2 interface. In experi-

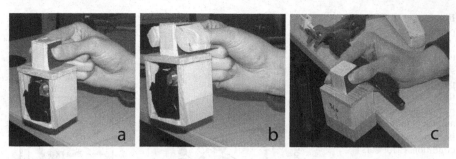

Fig. 3.9 Illustration of the three different conditions used in the experiment: (**a**) proprioception and cutaneous feedback (PC), (**b**) proprioception feedback, and (**c**) cutaneous feedback (C)

ment 2, we saw that this effect is not likely to be caused by the different grip forces that may be exerted during unimanual and bimanual weight lifting. In experiment 1 we also observed that weigh discrimination reduces dramatically with when lifting virtual weights generated with the MF2 device. It was suggested that this may be due to the lack of cutaneous cues. McCloskey [1] studied the contribution of cutaneous feedback but did not address the exclusive contribution of the proprioceptive cue. In experiment 3, we investigate the relative contribution of cutaneous and proprioceptive feedback on weight perception. We used a weight discrimination paradigm that tested sensitivity to weight changes using proprioception and cutaneous, proprioception-only or cutaneous-only feedback. The results have been previously reported in [23].

Methods

Apparatus and Stimulus   Nine weights ranging from 226–290 g were placed in nine wooden boxes weighted 42 g each providing an actual testing range from 268–332 g. Thus, the standard stimulus was 300 g and the step size 8 g. The top of the boxes was fitted with a grip block which was 23 mm wide (the grip aperture), 25 mm deep and 30 mm tall. The sides of the grip block were covered with a medium sandpaper (100 grade) to allow effective grip (Fig. 3.9).

Procedure   Three students (undergraduates and postgraduates) and two members of staff from the University of Birmingham, aged between 22 and 39, volunteered to take part in the study. The students were paid for their participation. Three conditions were tested: proprioception and cutaneous cues were available (PC), proprioception only (P) and cutaneous only cues were available (C) (Fig. 3.9). In the PC condition, the participants freely lifted the weights as in everyday activities. In the P condition the finger pads were covered with plastic cylindrical sections to prevent the use of cutaneous sheer forces caused by friction due to downward load forces. The participants lifted the weights in exactly the same way as in condition PC. In the C condition, a 'grip rest' was used to prevent the participants from using proprioceptive feedback either from the arm muscles or the finger joints. The grip was also padded to eliminate use of secondary pressure cutaneous cues.

Fig. 3.10 Overall performance in the three conditions and best-fit cumulative gaussian. The symbol • represents performance in PC condition, the symbol o represents performance in P condition and the symbol ▽ represents performance in the C condition. Error bars represent standard error

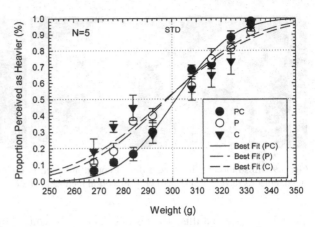

Results

The psychometric function (3.1) with a Gaussian rule to provide location and slope was used to fit individual and overall data. Here, performance in each condition was measured in terms of the DT defined as half the difference between the point at which 20% of the weights were perceived as lighter than the standard weight and the point at which 80% of the weights perceived as heavier than the standard weight; that is $DT = (DT_{80} - DT_{20})/2$.

Figure 3.10 shows the average performance under the three conditions. Performance in the PC condition was better than in P or C conditions which were very similar. Moreover, the standard deviations of the P and C Gaussians were 50% ($\sigma_p = 27.41$) and 67% ($\sigma_c = 30.43$), respectively, greater than the standard deviation of the PC function ($\sigma_{pc} = 18.27$).

This large differences in variability of performances between the condition PC and the conditions P and C may indicate that the two cues could be integrated on a basis of (MLE). The MLE rule states that the combined PC estimate has lower variance than the P or C estimate [24]. Since the variance is related to threshold, the relationship between the discrimination thresholds in the three different conditions is

$$DT_{pc}^2 = \frac{DT_p^2 DT_c^2}{DT_p^2 + DT_c^2} \tag{3.3}$$

Figure 3.11 shows the predicted discrimination threshold for the PC condition and the observed overall discrimination threshold for the conditions PC, P and C. In conditions P and C, discrimination thresholds tend to be higher than in PC indicating that a deterioration of performance when only the proprioception or cutaneous cues are available. In addition, it can be seen that the observed PC discrimination threshold is predicted well by the ME model.

Fig. 3.11 Predicted PC and observed PC, P and C discrimination thresholds based on 80% accurate judgements. The observe PC threshold is similar to the MLE prediction. Error bars represent standard error

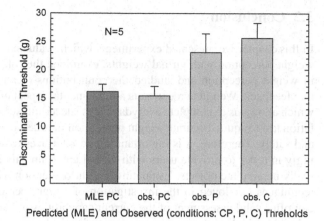

Predicted (MLE) and Observed (conditions: CP, P, C) Threholds

Discussion

Previous research has shown that proprioception and cutaneous cues seem to contribute to weight perception without assessing their relative importance [1]. Experiment 3 has provided additional evidence that restricting the proprioception or cutaneous feedback when lifting an object with precision grip tends to reduce sensitivity to weight changes. Discrimination thresholds tend to be lower when participants can have access to both proprioception and cutaneous weight signals than when either of the two is missing. Moreover, there is early indication that the integration of both cues may be based on a MLE rule. Even though there was no difference between the P and C conditions, participants reported that they found it more difficult to judge the difference of weights in the C condition. One plausible explanation could be the difference in the apparent weight between the C and P conditions. Since there was not lifting involved in the C condition, the apparent weight was approximately equal to the actual weight. However, in the P condition the apparent weight was a positive multiple of the actual weight since it was accelerating against gravity due to lifting forces. There is evidence that acceleration and deceleration during lifting may improve weight perception with real [25] and virtual weights [26]. Nonetheless, the results also indicate that absence or restriction of cutaneous feedback seems to result in a deterioration of weight sensitivity. This may be one of the reasons that in some applications of virtual weights loss or restriction of cutaneous feedback has been shown to decrease weight sensitivity by a factor of five as it was demonstrated in experiment 1 and [16, 27]. It is very likely that the acceleration and deceleration during lifting may affect the contribution of cutaneous feedback since skin deformation could increase due to greater sheer forces applied to stabilize a faster moving object.

3.2 Conclusion

In this chapter we presented experiments which evaluated unimanual and bimanual weight perception with virtual weights, examined the role of bimanual grip forces in weight perception and studied the contributions of cutaneous and proprioceptive feedback. Weight perception is a fundamental aspect of sensorimotor behaviour which allows us to complete everyday tasks successfully. In fact, all object manipulation tasks require accurate weight perception in order to carry them out effectively and safely. Therefore, it is important haptic interfaces to simulate weight convincingly in order to provide users with virtual environments in which they can effectively manipulate objects. Evaluating weight perception with virtual weights using techniques employed in the investigation of weight perception with real weights can offer valid assessment of the current capabilities of haptic interfaces as well as valuable guidelines for the improvement of their design.

Acknowledgements   This work was partly supported by the ImmerSence project within the 6th Framework Programme of the European Union, FET—Presence Initiative, contract number IST-2006-027141, see also www.immersence.info.

References

1. McCloskey, D.I.: Muscular and cutaneous mechanisms in the estimation of the weights of grasped objects. Neuropsychologia **12**(4), 513–520 (1974)
2. Brodie, E.E., Ross, H.E.: Sensorimotor mechanisms in weight discrimination. Atten. Percept. Psychophys. **36**(5), 477–481 (1984)
3. Flanagan, J.R., Wing, A.M., Allison, S., Spenceley, A.: Effects of surface texture on weight perception when lifting objects with a precision grip. Atten. Percept. Psychophys. **57**(3), 282–290 (1995)
4. Flanagan, J.R., Wing, A.M.: Effects of surface texture and grip force on the discrimination of hand-held loads. Atten. Percept. Psychophys. **59**(1), 111–118 (1997)
5. Flanagan, J.R., Bandomir, C.A.: Coming to grips with weight perception: Effects of grasp configuration on perceived heaviness. Atten. Percept. Psychophys. **62**(6), 1204–1219 (2000)
6. Gandevia, S.C., McCloskey, D.I.: Effects of related sensory inputs on motor performances in man studied through changes in perceived heaviness. Am. J. Physiol. **272**(3), 653 (1977)
7. Gandevia, S.C., McCloskey, D.I.: Sensations of heaviness. Brain **100**(2), 345 (1977)
8. Ross, H.E., Brodie, E.E., Benson, A.J.: Mass-discrimination in weightlessness and readaptation to Earth's gravity. Exp. Brain Res. **64**(2), 358–366 (1986)
9. Wing, A., Giachritsis, C., Roberts, R.: The power of touch: Weighting up the value of touch (2007)
10. Wall, S., Harwin, W.: Design of a multiple contact point haptic interface. In: Eurohaptics, Birmingham, UK, pp. 146–148 (2001)
11. McKnight, S., Melder, N., Barrow, A.L., Harwin, W.S., Wann, J.P.: Psychophysical size discrimination using multi-fingered haptic interfaces. In: Ser. Eurohaptics Conference, vol. 2004 (2004)
12. Michelitsch, G., Ruf, A., van Veen, H., van Erp, J.: Multi-finger haptic interaction within the miamm project. In: Proceedings of Eurohaptics, pp. 144–149 (2002)
13. Oyarzabal, M., Ferre, M., Cobos, S., Monroy, M., Barrio, J., Ortego, J.: Multi-finger haptic interface for collaborative tasks in virtual environments. In: Human-Computer Interaction. Interaction Platforms and Techniques, pp. 673–680 (2007)

14. Halabi, O., Kawasaki, H.: Five fingers haptic interface robot Hiro: Design, rendering, and applications. In: Zadeh, M.H. (ed.) Advances in Haptics (2010)
15. Giachritsis, C., Wing, A.: Unimanual and bimanual weight perception in a desktop environment. In: Ferre, M. (ed.) Proceedings of EuroHaptics (2008)
16. Giachritsis, C., Barrio, J., Ferre, M., Wing, A., Ortego, J.: Evaluation of weight perception during unimanual and bimanual manipulation of virtual objects. In: Third Joint. EuroHaptics Conference, 2009 and Symposium on Haptic Interfaces for Virtual Environment and Teleoperator Systems. World Haptics 2009, pp. 629–634. IEEE, New York (2009)
17. Gescheider, G.A.: Psychophysics: the Fundamentals. Lawrence Erlbaum, Hillsdale (1997)
18. Wichmann, F.A., Hill, N.J.: The psychometric function: I. Fitting, sampling, and goodness of fit. Atten. Percept. Psychophys. 63(8), 1293–1313 (2001)
19. Kjnoshita, H., Kawai, S., Ikuta, K.: Contributions and co-ordination of individual fingers in multiple finger prehension. Ergonomics 38(6), 1212–1230 (1995)
20. Kinoshita, H., Murase, T., Bandou, T.: Grip posture and forces during holding cylindrical objects with circular grips. Ergonomics 39(9), 1163–1176 (1996)
21. Gordon, A.M., Forssberg, H., Iwasaki, N.: Formation and lateralization of internal representations underlying motor commands during precision grip. Neuropsychologia 32(5), 555–568 (1994)
22. Giachritsis, C., Wing, A.: The effect of bimanual lifting on grip force and weight perception. In: Kappers, A., Van Erp, J., Bergman, T., Van Der Helm, F. (eds.) Haptics: Generating and Perceiving Tangible Sensations. LNCS, pp. 131–135. Springer, Berlin (2010)
23. Giachritsis, C., Wright, R., Wing, A.: The contribution of proprioceptive and cutaneous cues in weight perception: early evidence for maximum-likelihood integration. In: Kappers, A., Van Erp, J., Bergman, T., Van Der Helm, F. (eds.) Haptics: Generating and Perceiving Tangible Sensations. LNCS, pp. 11–16. Springer, Berlin (2010)
24. Ernst, M.O., Banks, M.S.: Human integrate visual and haptic information in a statistically optimal fashion. Nature 415, 429–433 (2002)
25. Brodie, E.E., Ross, H.E.: Jiggling a lifted weight does aid discrimination. Am. J. Psychol. 98(3), 469–471 (1985)
26. Hara, M., Higuchi, T., Yamagishi, T., Ashitaka, N., Huang, J., Yabuta, T.: Analysis of human weight perception for sudden weight changes during lifting task using a force display device. In: Proceedings of the 2007 IEEE International Conference on Robotics and Automation (2007)
27. Giachritsis, C.D., Ferre, M., Barrio, J., Wing, A.M.: Unimanual and bimanual weight perception of virtual objects with a new multi-finger haptic interface. Brain Res. Bull. 85(5), 271–275 (2011)

Chapter 4
The Image of Touch: Construction of Meaning and Task Performance in Virtual Environments

Miriam Reiner

Abstract One of the central questions in sensory interaction is how an image of the environment is constructed. In this chapter we report results of a series of studies conducted in a telesurgery system, aimed at exploring the underpinning cognitive process of construction of meaning through haptic dynamic patterns. We use a double analysis, the first rooted in behavioral methodology and the second rooted in signal processing methodologies, to validate the existence of a haptic language, and apply the results to challenging contexts in which haptic cues are on a subliminal level, in order to identify if and what is the role of subliminal cues in a haptic perceptual primitive language. We show that although participants are not aware of subliminal cues, nevertheless both action and perception are affected, suggesting that meaning is extracted from subliminal haptic patterns. Conclusion suggests a unified framework of a haptic language mechanism for construction of mental models of the environment, rooted in the embodied cognition, and enactive theoretical framework.

4.1 Introduction

This paper addresses two central questions related to understanding of task performance in immersive virtual environments. The first deals with the extraction of meaning from the touch patterns when interacting with the environment. The second takes a critical view and looks at conditions in which meaning, such as haptic properties, are hard to perceive such as when haptic cues are subliminal. In this second context we ask whether subliminal cues, which subjects are unaware of, contribute to the haptic cycle of action and extraction of information about the environment.

The first studies bring evidence of an underlying mechanism of haptic perception and meaning extraction. Results establish the concept of a 'haptic primitive language' related to the minimal cues needed to extract knowledge from a dynamic pattern of touch perception. Traditionally it is claimed the meaning is mainly in

M. Reiner (✉)
Department of Education in Technology and Science, Technion—Israel Institute of Technology, Haifa 32000, Israel
e-mail: Miriamr@tx.technion.ac.il

A. Peer, C.D. Giachritsis (eds.), *Immersive Multimodal Interactive Presence*, 51
Springer Series on Touch and Haptic Systems,
DOI 10.1007/978-1-4471-2754-3_4, © Springer-Verlag London Limited 2012

symbolic, 'propositional' representations, such as verbal or mathematical representations. We show here the meaning is not limited to symbolic representations and is inherent to perception of sensory haptic patterns.

This chapter will show that the cognitive system extracts meaning from haptic patterns, that meaning is in the dynamics of the input sensory cues, and that, similar to symbolic language, sentences and 'stories' can be constructed using haptic components that constitute a haptic primitive language. We show that a mental image of an object is constructed through integration of perceived changes over time of haptic properties, such as texture and compliance. Based on principles of correlational perception which suggest that perception is a statistical inference system, the haptic properties perceived are associated with memories of past experience, to generate the most probable gestalt image of an object.

The theoretical roots of the studies reported here are in the enaction approach to cognition. Enaction, first suggested by [1], opposes the central information processing view, and symbolic representations paradigm, and view cognition as grounded in the sensorimotor dynamics of the interactions between the human and the environment.

In the second part of the paper we apply the above principles to subliminal cues and ask how the perceptual construction of an image of an object through haptics is conducted when cues are subliminal. We ask how subliminal dynamic haptic patterns affect performance and decision making, and seek for the role of subliminal, unaware of, dynamic sensory input in a haptic primitive language. Stated differently, we ask whether haptic subliminal patterns, that occur during interaction with the environment, contribute to extraction of haptic information, and construction of a mental image of the environment. We conclude by integrating the results into an extended view of mechanisms for construction of images of the environment through haptic interaction.

4.2 Meaning and Perception of Dynamic Haptics

It is well documented that touching an object provides information about object properties such as hardness, compliance, shape, texture and similar. We suggest here that extraction of information is a dynamic rather than static process, involving motion of the palpating finger or hand, and that a mental image of an object is constructed through integration of all aspects of dynamic haptics-changes in compliance, in texture in curvature and similar.

This section looks at the link between the characteristics of the dynamic touch patterns and information. We ask what is the dynamic touch pattern that conveys particular information. It takes a cognitive perspective and identifies how conceptualization of the environment is related to haptic sensory input. For instance, when palpating the surface of human skin, what would one need to feel in order to say: I feel a fat/granular lump?

4.3 Goals and Methodology

The main goal of this study is then to construct a function that relates dynamic haptic patterns and corresponding meanings. This would be a 'reference system' since
it refers haptic patterns to specific interpretations. This constitutes a haptic primitive language, without syntax. Both terms, haptic-primitive-language, and reference-
system, are used here interchangeably.

In a haptically enhanced telesurgery system we identify re-occurring force-
sensation patterns while participants complete a haptic exploration task through
haptics only, then identify the meaning that subjects attach to the input haptic sensation patterns, and finally test the consistency and validity of the sensation-meaning
function.

4.4 Experimental Setup

Haptic patterns and associated meanings were explored in the context of a
telesurgery environment [2].

Subjects were asked to explore a remote silicon model of a breast in which three
'lumps' were inserted. The lumps varied in size, relative firmness and spatial location. Both transverse and normal feedback forces were recorded. Thus recording
showed the forces exerted on the hand while palpating the object using haptic devices which transferred the forces. The system is described in Fig. 4.1.

The system consists of two main parts—the operator's unit (1) and a remote performance unit (2) where the actual manipulation of objects happens. The subjects
sit at the operator's unit (A) looking into the virtual working space which is a 3D
strikingly realistic reproduction of the actual working space (B). Two physical handles are positioned in the virtual space. The operator grasps the handles and acts in
what appears to be the actual workplace. When the operator moves the handles, the
remote handles trace the operators movements. For instance, the operator can move
an object, observed in his working space but located further away. While moving it,
the operator sees and feels forces as if he touches the object directly. These forces
as well as the generation of the virtual working space are controlled by a computer.
The normal and tangential forces are recorded through a chart recorder. At the same
time subjects are videotaped while explaining what they think the object is. We create couples of force patterns and concepts mentioned, and identify the events when
each of the force patterns is coupled with the same concept.

4.5 Data

Data collected included both normal and tangential forces. The type of data collected is described in the Fig. 4.2. The upper signals describe the transverse feedback and lower describes the normal feedback forces.

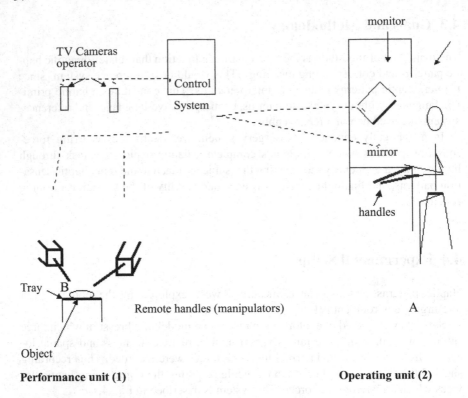

Fig. 4.1 Experimental setup

4.6 Analysis

While palpating subjects were requested to talk aloud [3] and describe their perception of the object palpated. Participants used descriptors such as: hard, firm, soft, smooth, edge (of a lump), circular, granular, round, elliptical. Two types of analysis were carried out. An obvious analysis is a wavelet or Fourier analysis. However, because of high levels of noise, we had to apply strategies to increase the signal to noise ratio, which we applied in the second type of analysis. In order to get an insight we decided also to run an analysis which is based on the fit between perception and haptic patterns. This is the first analysis. The second uses methods of signal processing.

The first analysis is statistical correlational: we identified all descriptors in the first three minutes of the palpation process. Then, the corresponding haptic patterns were identified. We focused on four descriptors:

1. Edge of a lump, normally associated with a large gradient in the force feedback appearing in the dynamic structure (see Fig. 4.3).
2. Sponginess, normally associated with non-organized high peaks in the force feedback.
3. Bumping into a metal tray, normally a single narrow peak in the force feedback;

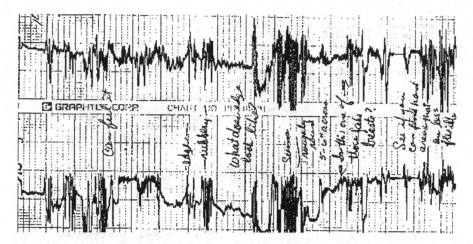

Fig. 4.2 The above is a segment of the haptic data collected. The *horizontal axis* is time. The upper signals describe the transverse feedback and lower describes the normal feedback forces. The written segments are descriptors used by the participants during the haptic exploration and were written in real time, in addition to recording of participants descriptions of what they felt ('think aloud' methodology)

4. A feel of a dip in the silicon model, normally starting with a local increase of the acceleration of the stylus, followed by low amplitude alternative force feedback;

Figure 4.3 describes the categories of patterns and associated descriptors. Using visual human analysis, we identified similar patterns to the above in the whole experiment: single peak (bumping into a hard metal), multiple non-organized peaks (spongy), sudden and long changes in force feedback, normally associated with a large gradient in firmness visible in the data as large changes in force feedback (edges of a lump), and a small change in gradient associated with a 'dip'. For validation reasons, identification has been done by three independent judges. Agreement between them was 0.99. Overall we identified 982 patterns. Then, we looked at the recorded data of the think aloud protocols, and identified the descriptors for each haptic pattern that we previously identified and validated. Finally, we calculated the fit between the descriptors and patterns: i.e. what is the percentage of patterns for which the descriptors were in the same category which was identified in the first iteration.

4.7 Results

Figure 4.3 brings examples of the descriptors used by three subjects, and a typical haptic pattern.

Figure 4.3 shows the patterns and the associated verbal representations. We plotted the ratio of the reoccurring patterns that are associated with the same meaning. Consistency is measured by the number of events in which the subject mentioned a

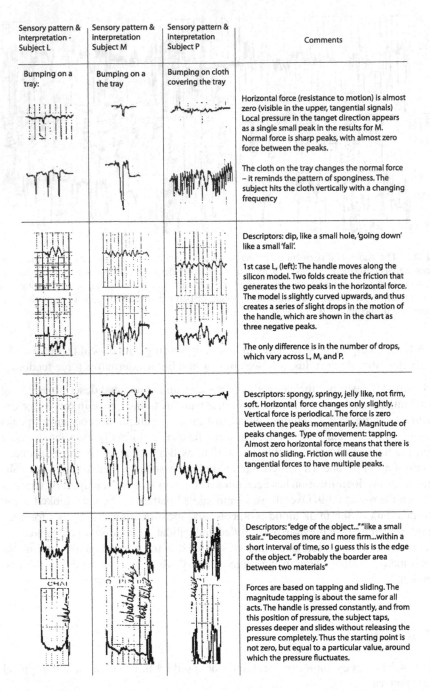

Sensory pattern & interpretation - Subject L	Sensory pattern & interpretation Subject M	Sensory pattern & interpretation Subject P	Comments
Bumping on a tray:	Bumping on a the tray	Bumping on cloth covering the tray	Horizontal force (resistance to motion) is almost zero (visible in the upper, tangential signals) Local pressure in the tanget direction appears as a single small peak in the results for M. Normal force is sharp peaks, with almost zero force between the peaks. The cloth on the tray changes the normal force – it reminds the pattern of sponginess. The subject hits the cloth vertically with a changing frequency
			Descriptors: dip, like a small hole, 'going down' like a small 'fall'. 1st case L, (left): The handle moves along the silicon model. Two folds create the friction that generates the two peaks in the horizontal force. The model is slightly curved upwards, and thus creates a series of slight drops in the motion of the handle, which are shown in the chart as three negative peaks. The only difference is in the number of drops, which vary across L, M, and P.
			Descriptors: spongy, springy, jelly like, not firm, soft. Horizontal force changes only slightly. Vertical force is periodical. The force is zero between the peaks momentarily. Magnitude of peaks changes. Type of movement: tapping. Almost zero horizontal force means that there is almost no sliding. Friction will cause the tangential forces to have multiple peaks.
			Descriptors: "edge of the object..." "like a small stair." "becomes more and more firm...within a short interval of time, so I guess this is the edge of the object." Probably the boarder area between two materials" Forces are based on tapping and sliding. The magnitude tapping is about the same for all acts. The handle is pressed constantly, and from this position of pressure, the subject taps, presses deeper and slides without releasing the pressure completely. Thus the starting point is not zero, but equal to a particular value, around which the pressure fluctuates.

Fig. 4.3 Description of intensity and change over time of tangent and normal forces exerted on the operator's handle. *Upper curve* in each cell is a display of tangential forces. *Lower curve* is of normal forces

similar verbal description for a similar force-sensation pattern. The following terms were correlated with distinct patterns (because of the high numbers, there is no additional gain from significance measures).

Edge—fit of 0.96 (variance: 0.001) between the patterns classified in the class of edge, and subjects' terminology describing an edge; Spongy—fit of (almost) 0.999 (variance: 0.001); Bumping into hard material—fit of (almost) 1; Dip, indention—fit of 0.85 (variance: 0.15).

The validity of the results of this study are limited—both because of the size of the sample and because of the limited analysis. For instance we have no evidence as to how consistent are these results for a wider variety of situations, subjects' background and cultures. Another issue is double meanings: how many times a particular pattern is coupled with more than one interpretation. There is no evidence concerning how learning of these patterns happens, what is the threshold force underneath subjects cannot identify forces, and what is the effect of scaling the forces. We do have evidence that shows that force sensations carry meaning, and are possibly rooted in an image schemata of correlational past sensory experience.

4.7.1 Quantitative Analysis: Extension of Data and Development of an Algorithm for Automatic Signal Processing of Haptic Exploration in a Remote Environment

Since the above results are limited, we performed an additional validation analysis, which allowed developing an algorithm for automatic signal processing of haptic exploration in a remote environment. In this analysis we suggest a method of processing of signals by looking at automatic detection changes of elastic-force feedback properties as a function of position in space and time. Since both time and spatial distances are relatively short, the signal-to-noise ratio for the data available is very low (the shorter the time duration of events the noisier the data received). The usual technique for signal processing may not produce accurate results. Thus, in order to avoid the low SNR, we apply a method of processing that can possibly extend the statistics of the data available. The main idea of this method is to expand the set of the data to a set of Planar Graphs. Integral Geometry Transformation is applied to the existing data to receive a new Characteristics Function, which contains more information about the statistics of data available and still contains all the peculiarities of the former.

The data of values available is considered a set of samples received from the haptic measures—a two channel time series data. We can write it as a function $s = s(t)$ for each channel. For two channel data we can write $s1 = s1(t)$, $s2 = s2(t)$, where t is time and $s1(t)$, $s2(t)$ are the processed data. We consider this as a set: $c1 = u1(t, s1(t))$ and $c2 = u2(t, s2(t))$ where $c1 \equiv c2 \equiv 1$ are constants.

The multi-channel time signal can be regarded as a set of curves in the Euclidean plane

$$\left\{ c = u_i\left(t, s_i(t)\right) \right\}_{i=1}^{i=N} \tag{4.1}$$

where N is the number of channels, $s_i(t)$ are the continuous data functions of the channels, t represents a function argument and c are constants.

Thus, we produced an extension of the data dimension from **R** (real number axis) to E^2. This extension allows additional information about data and improved signal-to-noise ratio in further more traditional processing.

The Cauchy theorem claims that 'Curve length is the average length of the orthogonal projection of the given curve on all lines through the origin', and hence allows analysis of the integrated set of projections, without distortion of the characteristics of the results. The integrated set of projections is an extended data set that reflects the same patterns as the original, with improved SNR. Classification of the patterns according to frequency and amplitude, suggest an almost identical classification to the previous, which did not even require statistics, since it simply overlapped with the original classification. For example, Fig. 4.4 shows a segment of the haptic signals. The red, pink and green lines show the different patterns identified by the latter methodology. The black lines show the previous analysis, based on human, naked eye, classification and correlations between dynamic haptic patterns and verbal descriptors, expressed by participants in the think-aloud process. The overlap in this segment, shows that not even one haptic pattern was missed by the automatic algorithm. This reflects the results in all obtained haptic data.

To summarize: we used two types of analysis to establish the concept of a haptic reference system, i.e. a primitive haptic 'language'. The results show that the cognitive system extracts meaning from haptic patterns, that meaning is in the dynamics of the input sensory cues, and that, dynamics over time and space determines the information conveyed through haptics. We found that it is not a single point of haptic interaction, which suggests the attached meaning but rather integration over time and space of haptic cues. Figure 4.4 shows the time order of exploration: sharp gradient of compliance is not sufficient to extract a shape. However, we found that a perceived continuity in space of such a gradient, suggests the existence of contours of an object. Integration of cues over time suggests a gestalt image of an object. Similar to symbolic language, sentences and 'stories' can be constructed using haptic components that constitute primitive haptic language. Whether this primitive language has also hidden grammatical rules that may account for a syntax requires additional research.

4.7.2 Embodied Cognition and the Emergence of a Haptic Primitive Language

The previous section established the idea that meaning is embedded in haptic dynamic patterns. Cognitive processes of meaning extraction are not just in words and symbolic systems, but rooted in the sensory interaction with the environment. Perception-of, and action-on the world are shaped by both the mind and the environment suggesting that the environment has a dynamic active role in cognitive processes. According to [4], the basic notion of embodiment is generally understood as the way an organism's sensorimotor capacities enable it to successfully inter-

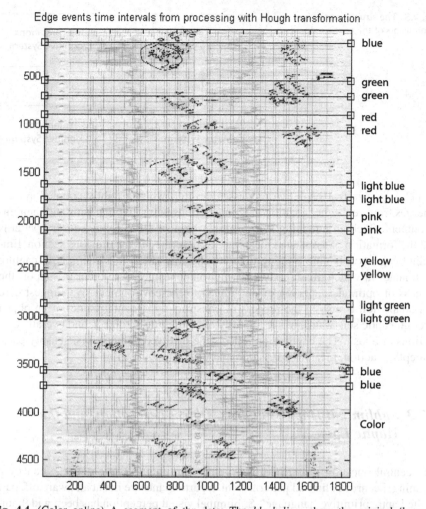

Fig. 4.4 (Color online) A segment of the data: The *black lines* show the original 'human eye'-based analysis. The *color lines* show the analysis of the automatic algorithm. The colors suggest the classification, i.e. the meaning. The original chart is brought here, to show how the data was collected: visual description of the changes in haptics, and the verbal descriptors, used by the subjects, and written by the experimenter. These written descriptors served later on as clues to help making sense of the match between the haptic patterns and the verbal descriptors used by the subjects

act with its environment. Interaction with the environment is not limited to motor learning, and has impact on the higher reasoning faculties. Cognition is embodied in the sense that some of the cognitive features depend upon characteristics of the human physical body. The patterns of sensory interactions become memories that affect future behavior and casual modeling [4, 5].The haptic-meaning association suggests a model of how embodied sensory cues are linked to higher cognitive faculties of modeling the world. The general embodied cognition theory first coined

Fig. 4.5 The structure and components of the haptic loop

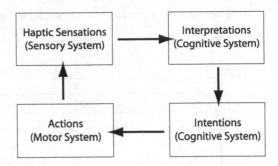

by [1], suggests that cognitive processes develop when a tightly coupled system emerges from real-time, goal-directed interactions between organisms and their environment. Such an emerging system might be a set of dynamic sensory patterns and the formation of associated meanings, as described in the previous section. Embodied cognition is thus framing theoretically this chapter: it attempts to capture the manner in which mind, body, and environment interact in order to support the process of optimal adaptation of the human to the environment. We suggest here that adaptation is also in the process of constructing a mental model of an object through haptic sensations, of successfully manipulating an object, and overall completions of a task. The haptic loop reflects the involved components: sensory cues, perception, action, meaning extraction and goal setting, see Fig. 4.5.

4.7.3 Subliminal Haptic Patterns, Object Perception, and the Haptic Loop

The central question raised here is whether subliminal haptic patterns convey a meaning, i.e. are subliminal patterns, which participants are not at all aware off, part of the haptic primitive 'language'. Subliminal visual perception has been widely investigated (e.g. [6]). So are subliminal auditory perception (e.g. [7, 8]). They show that visual cues, under the threshold of perception indeed affect perception. In this study we ask what is the impact of haptic stimuli on both action and cognition. Two studies were conducted ([9, 10]). The first looked at the impact of subliminal haptic cues on the haptic loop (Fig. 4.5). Methodology was based on [11] and [12]. Participants held a handle of a phantom in a collocated stereographic virtual world. The phantom was represented in the virtual world by a stylus. The virtual world included a surface and a scale (Fig. 4.6). Subjects were asked to glide the stylus on the surface. At a randomly set point on the surface, roughness changed by a value that was either above threshold, exactly threshold or lower than the threshold of roughness perception. Individual thresholds were measured prior to the experiment. Results suggest that participants tend to apply a larger force on the surface when the roughness is increased even if the change was below the threshold, and participants claimed that the surface was uniform. This suggests that haptic subliminal cues have an impact

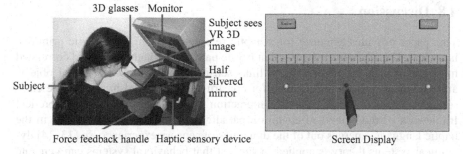

3D glasses   Monitor

Subject sees VR 3D image

Half silvered mirror

Subject

Force feedback handle   Haptic sensory device

Screen Display

Fig. 4.6 The visuo-haptic virtual world, the surface and the stylus

Fig. 4.7 Participants were asked to glide the stylus on two surfaces that differ by subliminal level of roughness

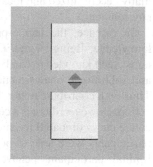

on the manual action response. This further suggests that subliminal cues are part of the reservoir of haptic language, in spite of participants being unaware of these type of haptic cues. These results suggest that subliminal cues affect the haptic-action component in the haptic loop. However, the haptic loop includes also a cognitive component. Thus for the subliminal cues to be part of a haptic reference system, it is essential to show that subliminal cues affect also cognitive processes. In an additional study we used a similar virtual haptics-enabled virtual reality to test cognitive effects of subliminal haptic cues. Participants were asked to glide the stylus on two surfaces that differ by a subliminal value of roughness (Fig. 4.7). The individual thresholds were tested prior to the main experiment. Then, participants were asked whether the two surfaces differ. Results show that participants did not recognize any difference between the two surfaces. However, when forced to chose one of the surfaces, most participants chose smoother surfaces, in spite claiming that there is no difference between the two surfaces (for details, please see [9]). Conclusion of this experiment suggest that haptic subliminal cues convey a meaning: when forced to chose, participants seem to 'know', in an embodied but not verbal sense, which surface was smoother/rougher. Thus subliminal cues, just as supraliminal cues, convey a meaning, and contribute to construction of an image of the environment.

4.8 Discussion

This paper provides evidence for a haptic reference system, or a haptic primitive language. The claim made here is that haptic patterns are memorized and correlated meanings. Correlational perception links sensory patterns with images of objects and provide a possible mechanism for construction of mental images of the environment. The central role of bodily interaction matches and well fits the theoretical framework of the embodied cognition paradigm. According to this paradigm the haptic language emerges out of the dynamics of sensory experience. In [13, 14] dynamical systems theory is applied to suggest that behavioral systems can generate novel behaviors such as different paths for optimal reaching objects or different patterns of goal-oriented manipulation of objects. Such novel behaviors evolve through ongoing bodily action. Haptic language can be viewed as self organizing product of many decentralized and local interactions taking place in bodily interaction with the environment.

In essence, the haptic-primitive-language system is a mechanism to extract knowledge. Being rooted in the sensory-bodily system rather than symbolic systems, such knowledge can be considered embodied. Both supraliminal, aware cues, and subliminal, unaware cues were found to affect motor and cognitive behavior. Thus both contribute to the 'knowledge' acquired by the system from sensory interaction. However, there is a central difference between the knowledge gained through aware cues compared to unaware cues: the first is explicit knowledge, for instance, participants were able to point at the rougher area. However, when cues are subliminal, participants behaved as if the know the difference in texture, but when asked that did not reflect such knowledge, thus knowledge is implicit. The haptic-primitive-language system is then a mechanism for abstraction of knowledge about the environment on two levels: explicit and implicit, each being activated, through adaptation processes, according to the environmental sensory cues.

A closely related view that supports the existence of a haptic language is that of enactive knowledge. Enactive knowledge is information gained through perception-action interaction in the environment. It emerges from action and is constructed on motor skills, such as manipulating objects, riding a bicycle, cooking, or manipulating the complex sensory input in playing basketball. Enactive knowledges of entities are those that emerge out of doing. The applied aspects of enaction are in the construction of haptic interfaces, that through 'record' of patterns of forces are able to re-play these patterns to convey the natural meaning. For instance recording of the haptic patterns of interaction with a silicon breast model, might be re-played in a virtual world, to convey the valid experience and meaning of a virtual breast.

This study reported here was mainly dedicated to haptic language, but haptic pattern perception is enhanced by coordinated multimodal cues, as shown by Hecht and colleagues ([15, 16]). In fact, it has been shown that haptic perception is processed in the visual system, in an area known as LOtv: Lateral Occipital tactile visual area, in addition to motor areas, suggesting that haptic language may rely on multimodal cues ([17–19]). Comparison between neural correlates of multisensory cues, relative to unisensory cues, showed a clear speed of processing advantage to multisensory cues [20]. Whether haptic language is a subset of a sensory non-propositional, non-symbolic mechanism of construction of 'presence', i.e. a mental model of the environment, might be feasible, yet still need to be proven.

Acknowledgements   This work was partly supported by the ImmerSence project within the 6th Framework Programme of the European Union, FET—Presence Initiative, contract number IST-2006-027141, see also www.immersence.info.

References

1. Varela, F.J., Thompson, E., Rosch, E.: The Embodied Mind: Cognitive Science and Human Experience. MIT Press, Cambridge (1991)
2. Green, P.S., Hill, J.W., Jensen, J.F., Shah, A.: Telepresence surgery. IEEE Eng. Med. Biol. Mag. **14**(3), 233–261 (1995)
3. Newell, A., Simon, H.A.: Human Problem Solving. Englewood Cliffs, Prentice-Hall (1972)
4. Shapiro, L.: Embodied Cognition. Routledge, New York (2011)
5. Clark, A.: Supersizing the Mind: Embodiment, Action, and Cognitive Extension. Oxford University Press, New York (2008)
6. Reingold, E.M., Merikle, P.M.: Using direct and indirect measures to study perception without awareness. Atten. Percept. Psychophys. **44**(6), 563–575 (1988)
7. Groeger, J.A.: Qualitatively different effects of undetected and unidentified auditory primes. Q. J. Exp. Psychol., A Hum. Exp. Psychol. **40**(2), 323–339 (1988)
8. Suzuki, M., Gyoba, J.: Visual and cross-modal mere exposure effects. Cogn. Emot. **22**(1), 147–154 (2008)
9. Hilsenrat, M., Reiner, M.: The impact of subliminal haptic perception on the preference discrimination of roughness and compliance. Brain Res. Bull. **85**(5), 267–270 (2011)
10. Hilsenrat, M., Reiner, M.: The impact of unaware perception on bodily interaction in virtual reality environments. Presence: Teleoperators and Virtual Environments **18**(6), 413–420 (2009)
11. Zajonc, R.B.: Attitudinal effects of mere exposure. J. Pers. Soc. Psychol. **9**(2p2), 1 (1968)
12. Kunst-Wilson, W.R., Zajonc, R.B.: Affective discrimination of stimuli that cannot be recognized. Science **207**(4430), 557 (1988)
13. Thelen, E., Smith, L.B.: A Dynamic Systems Approach to the Development of Cognition and Action. MIT Press, Cambridge (1994)
14. Thelen, E., Smith, L.B.: Development as a dynamic system. Trends Cogn. Sci. **7**(8), 343–348 (2003)
15. Hecht, D., Reiner, M., Karni, A.: Multisensory enhancement: gains in choice and in simple response times. Exp. Brain Res. **189**(2), 133–143 (2008)
16. Hecht, D., Reiner, M., Karni, A.: Repetition priming for multisensory stimuli: Task-irrelevant and task-relevant stimuli are associated if semantically related but with no advantage over uni-sensory stimuli. Brain Res. **1251**, 236–244 (2009)
17. Neuper, C., Scherer, R., Reiner, M., Pfurtscheller, G.: Imagery of motor actions: Differential effects of kinesthetic and visual-motor mode of imagery in single-trial EEG. Cogn. Brain Res. **25**(3), 668–677 (2005)
18. Reiner, M., Stylianou-Korsnes, M., Glover, G., Hugdahl, K., Feldman, M.W.: Seeing shapes and hearing textures: Two neural categories of touch. Open Neurosci. J. **5**, 8–15 (2011)
19. Stylianou-Korsnes, M., Reiner, M., Magnussen, S.J., Feldman, M.W.: Visual recognition of shapes and textures: An fmri study. Brain Struct. Funct. **214**(4), 355–359 (2010)
20. Birnboim, I.: Early cortical interactions in multimodal perception. PhD thesis (2009)

Chapter 5
Psychological Experiments in Haptic Collaboration Research

Raphaela Groten, Daniela Feth, Angelika Peer, and Martin Buss

Abstract This chapter discusses the role of psychological experiments when designing artificial partners for haptic collaboration tasks, such as joint object manipulation. After an introduction, which motivates this line of research and provides according definitions, the first part of this chapter presents theoretical considerations on psychological experiments in the general design process of interactive artificial partners. Furthermore, challenges related to the specific research interest in haptic collaboration are introduced. Next, two concrete examples of psychological experiments are given: (a) A study where we examine whether dominance behavior depends on the interacting partner. The obtained results are discussed in relation to design guidelines for artificial partners. (b) An experiment which focuses on the evaluation of different configurations of the implemented artificial partner. Again the focus is on experimental challenges, especially measurements. In the conclusion the two roles of experiments in the design process of artificial partners for haptic tasks are contrasted.

5.1 Introduction

While haptic interaction with real or virtual objects has been researched intensively in haptics literature, haptic collaboration taking place in human-human or human-robot dyads attracted less attention. However, the number of relevant applications increased in recent years asking for a profound investigation of this topic. In order to define this research topic, we have to consider that "whereas interaction entails

R. Groten (✉) · D. Feth · A. Peer · M. Buss
Institute of Automatic Control Engineering, Technische Universität München, Munich, Germany
e-mail: r.groten@tum.de

D. Feth
e-mail: daniela.feth@tum.de

A. Peer
e-mail: angelika.peer@tum.de

M. Buss
e-mail: mb@tum.de

A. Peer, C.D. Giachritsis (eds.), *Immersive Multimodal Interactive Presence*,
Springer Series on Touch and Haptic Systems,
DOI 10.1007/978-1-4471-2754-3_5, © Springer-Verlag London Limited 2012

action on someone or something else, collaboration is inherently working *with* others" [1] (referring to [2, 3]) and consequently requires sharing of goals. This implies the recognition of the partner's intentions (i.e. action plans towards a goal) and their integration into one's own intentions. Since shared intentions "are not reducible to mere summation of individual intentions" [4], two collaborating systems share at least one goal and are confronted with the challenge to find suitable action plans to achieve it [5–7]. Haptic collaboration is based on the exchange of *force and motion signals* between partners, either in direct contact or via an object. As long as there is physical contact between the two partners, the physical coupling leads to a constant signal flow between them. Thus, haptic collaboration is *simultaneous and continuous* because the partner's dynamics are perceived while acting. This instantaneous feedback is the main difference to turn-taking in talking and to forms of collaboration, where cooperation takes place mostly sequentially, see e.g. [8–11].

There is a wide application area of haptic collaboration. Here we list some of them without assuming completeness: (i) industrial applications, where bulky and heavy objects need to be manipulated in collaboration with robots; (ii) assistant robots such as walking helpers robots providing support to human; (iii) motor-rehabilitation, where human limbs are guided by a robotic system along predefined trajectories; (iv) motor skill learning, where a complex motor skill is transferred from experts to novices [12–15]; (v) teleoperation, where the execution of tasks in remote or virtual settings is supported; (vi) social interaction like dancing or handshaking (see Chap. 11) where a virtual partner or robot takes over the role of one human partner.

This large variety of applications motivates the design of artificial partners that are able to not only interact, but also to collaborate with humans in performing haptic tasks. In developing such artificial collaborative partners psychological experiments play an important role. After a short description of psychological experiments, their importance in the design process of interactive artificial collaborative partners is outlined and challenges in designing and conducting psychological experiments in haptic collaboration research are discussed. Two concrete examples located in early and late design stages are finally used to illustrate the importance of psychological experiments in the development cycle of artificial collaboration partners.

5.1.1 Definition of Psychological Experiment

Wundt was the first scientist, who stressed the meaning of experiments in psychological research [16]. In general terms, psychological experiments can be defined as follows: "In an experiment, scientists manipulate one or more factors and observe the effects of this manipulation on behavior" [17]. The manipulated factors or independent variables and their levels define the experimental conditions. Their effect is assessed by measures (of behavior). These measures are also termed dependent variables. Because of the complexity of human behavior, the experimenter needs to set up the experimental conditions carefully in order to ensure that any effect on measurements is caused by a systematic variation of the independent variable.

When conducting psychological experiments the following steps are undertaken in line with the definition of experiments (compare e.g. [18]):

- Intentional preparation and selection of experimental conditions
- Control for unsystematic influences and differences between participants
- Systematic variation of experimental conditions
- Observation of effects due to variations in experimental conditions in measures/dependent variables

Now that a general idea of psychological experiments is given, the next section discusses their role when designing interactive systems.

5.1.2 Psychological Experiments in Design Processes

Design processes for interactive systems can generally be described by the following four stages (compare e.g. [19, 20]; ISO 9241-210 (former ISO 13407)):

1. Identification of requirements
2. Design
3. Development
4. Evaluation

In the following paragraph these stages are interpreted for the collaboration between a human and a virtual or robotic collaborative partner: A fundamental stage is to identify requirements, which have to be met by the partner. This implies not only an understanding of the task/scenario, but a profound knowledge about the human collaborator. Possibilities to integrate this knowledge in the control architecture of artificial collaboration partners are considered in the second stage. Afterwards a first implementation of the artificial partner can be developed. How good the actual performance matches the formulated requirements is investigated in the last stage. Depending on achieved evaluation results the whole process is repeated iteratively.

While the actual development, implementation, and technological evaluation of artificial collaboration partners belongs to the science of engineering and computer science, user-centered evaluations require psychological methods. In this chapter we not only illustrate how psychological experiments in the last stage of the design process can contribute to the user-centred evaluation and comparison of different partner implementations, but also emphasize their importance for early design stages that aim at formulating requirements.

Experiments conducted in early phases and accordingly gain knowledge can influence the design of haptic collaboration partners. Such experiments have the potential of saving time and costs by preventing disadvantageous implementations, which can otherwise only be detected when the implementation is finished. In addition, experiments serving as an evaluation method, allow insights in the success of ongoing implementations and design decisions. Thus, they allow addressing user acceptance and efficiency of the implementation before it is further developed and finalized.

5.2 Challenges for Psychological Experiments in Haptic Collaboration

Before we present two concrete examples of psychological experiments that have been conducted to provide design guidelines and evaluate artificial partners, this section discusses general challenges when setting up experiments in the field of haptic collaboration.

5.2.1 Approach

In line with the aforementioned two different roles of psychological experiments in the design process of interactive systems, we distinguish two different approaches to generate knowledge on haptic collaboration:

1. *Investigation of human-human collaboration*: The first approach considers studying two collaborating humans with the goal of knowledge-acquisition on underlying principles of haptic collaboration. The main aim is to study "natural" collaboration (in contrast to newly learned artificial signals, i.e. haptic icons) and to derive models and to investigate patterns that help to enhance the design of artificial partners. Of special interest is the derivation of behavioral models explaining the behavior of *one* human partner in the dyad. The implementation of such models on a robot—one human of the collaborating dyad is "substituted" by a robot—allows the collaboration with a artificial partner that shows human-like behavior. This approach is located early in the design process as it defines requirements of artificial partners *before* they are developed.
2. *Direct investigation of human-artificial partner collaboration*: The second approach considers investigating directly how humans collaborate with artificial partners. One main interest are evaluation results on specific parameters values of the artificial partner's control architecture. This approach, however, requires an existing model of the collaboration partner and is consequently chosen in a later stage of the design process, *after*—based on pre-knowledge—a first prototype of the artificial partner has been built.

Hence, the first challenge when designing experiments for haptic collaboration lies in a clear definition of the role of the experiment in the design process. It is desirable to combine both approaches when building artificial collaboration partners. Starting with the first approach allows to gain information on key-concepts and influencing factors and to come up with a first model of the collaboration partner.

Once a model is found and implemented, it needs to be evaluated. Running through this process in an iterative manner allows the refinement of artificial collaboration partners.

Currently, only very few models of haptic collaboration partners are available that can be implemented. Thus, recent research in this field relies on using haptic human-human collaboration as a reference when designing artificial partners, i.e. follows the first approach [21–24]. Authors following these approach argue that intuitive haptic collaboration between humans and artificial partners has to be based on rules and models familiar to humans. One strong argument in favor of this approach is that if no other mental model[1] is available, we tend to use a mental model of ourselves to predict the behavior of our collaboration partner [28].

5.2.2 Determining Factors and Control Conditions

After the approach is clarified, the actual experiment needs to be designed. Depending on whether the experiment has an exploratory character or is testing hypotheses in relation to an existing theory of human behavior, the first step lies in the definition of a research question or hypothesis. As mentioned in the general description of experiments, a research question addresses the effect of a controlled change in a factor on a selected measurement. In the following paragraphs we briefly review some factors that are of relevance for a deeper understanding of haptic collaboration partners.

First, studies investigating haptic collaboration focused mainly on the effect of mutual haptic feedback, striving to understand how the haptic communication channel between partners is influencing their behavior. In order to understand this effect, a condition where haptic collaboration can take place, i.e. mutual haptic feedback between partners is provided, has to be contrasted to a control condition where specific characteristics of haptic collaboration are not given. Thus, the difference between those conditions can be explored in measures. In dependence of the introduced control conditions, the conclusions, which can be drawn from differences between the experimental conditions, vary. The following control conditions can be distinguished:

(1) *Single-person, single-hand control condition*: In this condition, see e.g. [29, 30], collaboration does not take place by definition. Mental models of the partner and coordination of two individually applied forces are not necessary. Hence differences between a single-person condition and a haptic collaboration condition involving two collaboration partners can have several reasons, i.e. the effect of the haptic feedback, the increased workload due to action plan integration, the

---

[1] Mental models are an internal representations of the external world, including the collaborating partner. Mental models allow to explain and predict a system state and to recognize the relationship between system components and events [25]. Recently mental models receive increasing attention in interaction design processes [20, 26, 27].

task simplification due to the support of the partner, and possible social effects (to name some sources of variations in measurements). This implies that conclusions drawn related to this control condition are challenging to interpret.

(2) *Single-person, dual-hand control condition*: One idea to overcome the problems of the single-person, single-hand control condition is to introduce a dual-hand condition. The dual-hand condition does not require building mental models of the partner, but requires coordination of the two hands. The dual-hand condition can be presented with and without haptic feedback, and thus, allows studying the effect of feedback separately from the effect of interaction on motor-coordination. The effect of coordination with a partner can be examined when comparing this control condition to the experimental one. The challenge here is that the single person has only one dominant hand, whereas partners in a dyad can both work with their dominant hands. Thus, the interpretation of differences in task execution between those two conditions is again challenged.

(3) *Two partners, without-haptic-feedback control condition*: This control condition requires building of mental models on the physical coordination and cognitive level, which is also common to the haptic condition. Providing visual feedback from the partner's actions only, however, potentially leads to inconsistencies when two persons jointly manipulate an object. For the individual the proprioceptive movement of the muscles and the so-estimated object movement is not necessarily consistent with the real object movement, which is also influenced by the partner. Thus, this control condition confounds the effect of additional haptic information from the partner with effects due to disturbances related to this feedback. In addition, two cases have to be distinguished: (a) the haptic feedback of the object is still provided or (b) no haptic feedback at all is given (e.g. [31, 32]). In the latter case, the overall effect of haptic feedback is investigated, which is of special interest in VR applications. If the goal is to address the effect of haptic feedback on the actual collaboration, thus communication between partners, the first condition should be used. However, the challenge underlying this condition lies in the decision to determine how the haptic feedback from the object should look like, as this feedback will no longer have an equivalent in real life. This decision can again influence found effects in observed measures.

(4) *Artificial partner*: Comparing a artificial partner with a human partner is foremost done to evaluate a model of a haptic collaboration partner. Differences between the two conditions allow defining the quality of such a model. Because the model needs to be defined beforehand, this control condition is added for the sake of completeness, but has only limited use as the state of the art does not provide advanced artificial partners.

Summarizing we can state that all introduced control conditions have their advantages and disadvantages when interpreting differences between them and two haptically collaborating human partners. In the next section we will address challenges, which lie in the measures themselves.

5.2.3 *Measures*

The goal of this section is not to introduce specific measures relevant for haptic collaboration research. Instead, a general overview on challenges is provided. In literature different groups of measures used for the investigation of haptic collaboration
can be distinguished:

(1) *Task performance measures*: The precise formulation of such measures depends
    highly on the task, which is executed in a specific scenario. Performance mea-
    sures can be time-related, e.g. time to task completion [33], error-related, e.g.
    dropping of the jointly manipulated object [32, 34], or position-related when
    the deviation from a desired trajectory is measured, e.g. [35–37]. Since in hap-
    tic collaboration the two partners act towards a shared goal, task performance
    can often be measured for the collaborating dyad only and not for each individ-
    ual separately.

(2) *Force- and energy-related measures*: When two partners manipulate a joint ob-
    ject or guide each other, forces and energy are exchanged between them (and the
    object), compare e.g. [37]. Interaction forces can be measured with force-torque
    sensors and energy calculated by additionally taking the position information
    into account. Without considering different force and energy measures in detail,
    the following challenges can be summarized:

    (a) Measures in haptic collaboration are based on physical variables, which
        results into *time-series data* when collected. Specific information (param-
        eters) has to be extracted to make interpretation possible. This can be
        achieved by means of statistical analysis, time series analysis, or dynamic
        modeling.

    (b) It is important to differentiate between data representing interacting *indi-
        viduals* and *dyads*. Depending on the analysis level, modeling assumptions
        have to be checked (i.e. two individual data streams within a dyad are not in-
        dependent) and conclusions of data analysis have to be adapted to the level
        of the unit of analysis. The individual behavior within a dyad is the data
        most interesting for the implementation of a artificial collaboration partner
        that behaves human-like. However, it is also the data most difficult to in-
        vestigate as standard procedures of inference statistics cannot be applied as
        detailed in the next section.

    (c) Measures of forces in haptic collaboration comprise challenges in relation
        to the interpretation of the cause of the measured force and the *responsi-
        ble partner*. In line with [38], measured forces can be actively applied or
        result from passive behavior, e.g. when partner 1 is pulling an object not
        only her/his forces are recorded, but also forces applied by the other partner
        are measured. This partner might apply forces actively by injecting energy
        into the system or passively by using his arm inertia to resist the partners
        motions.

(3) *Questionnaire-based measures*: These measures are not addressing behavior as
    they are mostly evaluated after the task has been executed. Instead, their advan-
    tage lies in the possibility to explore subjective user experiences, which are not

directly accessible in behavior as e.g. the felt success of collaboration or co-presence [39]. A general drawback of questionnaire data in contrast to online gained behavioral measures is that questionnaires tend to be less reliable. This means that due to the subjectivity and complexity of the measured concepts, there can be a lot of noise added to the most representative value/answer.

Adequate measures are of importance in order to answer specific research questions and have to be developed in the given context. In the next section we will discuss aspects of the analysis of the so-gained haptic collaboration data.

5.2.4 Analyzing Interactive Experiments

When conducting experiments in collaborative settings, where more than one human is involved, it needs to be decided, which participant's behavior is analyzed. When analyzing one partner, the challenge lies in the fact that standard inference statistical methods require independence of data in one condition. This independence is not given if we collect data from interacting partners and then analyze the two individuals, or if one participant is interacting with more than one partner: If two partners are physically coupled and the reaction of one partner is influenced by the action of the other, independence is threatened when e.g. considering position data, but additionally cannot be assumed for higher level measures, as e.g. perceived quality of collaboration. In social psychology general experimental designs and according methods of analysis have been introduced [40–47].[2] We can distinguish the following solutions to the problem of interactive data analysis:

1. The dyads taking part in the experiment are *independent*, i.e. no partner is part of more than one dyad. Further, we analyze the data on the *dyadic level*, e.g. calculate the mean from the two individual measures. However, individual behavior within an interacting dyad cannot be addressed in this way.
2. Another approach when examine *individual* behavior is to standardize one partner, i.e. to employ a *confederate* and to assume his/her behavior to be constant in all interactions. Then, only the second partner, the participant, is analyzed. The drawback of this procedure is that interaction involves adaptation towards the partner. However, the trained partner cannot adapt naturally to the partner as his/her goal is to present a standardized partner.
3. A third possibility to deal with the question of how to examining behavior in interaction is to directly address it. This can be done by using a more advanced experimental design, which allows participants to interact with several partners (e.g. round robin design [47]). This design requires an analysis which is directly modeling the interdependence of data gained by two interacting participants. An example from social psychology will be given in the next section.

---

[2]For a general overview on experimental design and statistical analysis please refer to [17, 48–55].

Next to inference statistic methods, dynamic modeling of behavior plays an important role when describing human behavior in haptic collaboration as dynamic models can be potentially implemented on robotic partners. In any case, the question arises how representative parameters describing human behavior can be found. Next to the possibilities offered by statistics e.g. in the form of confidence intervals, the representativity of the sample itself is important. Most of all, this implies an appropriate sample size.

Following these general comments, the next two sections will introduce two examples of experiments conducted in the context of haptic collaboration. Each experiment is addressing a different stage in the design process of haptic collaboration partners: The first experiment focuses on an early design stage which aims at formulating requirements, the second experiment focuses on a late design stage which aims at evaluating and comparing existing implementations.

5.3 Example 1—Early Design Stage: Identification of Requirements

In this subsection, we will introduce a psychological experiment, which aims at understanding generic principals of haptic collaboration. The experiment is located within the first design step for interactive systems and will investigate two collaborating humans. The study presented in the following has been published as part of [56], where more details on the state of the art, measurements and results can be found.

5.3.1 Research Interest

Each of the collaborating partners in jointly executed haptic tasks is only partly responsible for the resulting behavior of the overall system and consequently contributes only partly to the overall task performance (shared action plans). When sharing the responsibility of the task outcome, partners can decide for different distributions of workload among them, which results into different patterns of dominance.

Dominance "refers to context- and relationship-dependent interaction patterns in which one actor's assertion of control is met by acquiescence from another" [57] and is described as "a relational, behavioral, and interaction state that reflects the actual achievement of influence or control over another via communicative actions" [58].

Of special interest in this context is whether human individuals have a tendency to show the same dominance behavior with all partners they collaborate with or whether their role is created newly with each collaboration partner. Hence, we raised the following research question:

How consistent is (mean) dominance behavior across several partners and subtrials?

Information on this subject allows insights on the required adaptation of a artificial collaboration partner towards a specific human partner.

5.3.2 Experimental Approach

We decided for a standardized experiment with high internal validity when investigating the role of mutual haptic feedback and dominance behavior. In a first step an appropriate task had to be found. One assumption underlying the design of our experiment is that most tasks which require haptic collaboration can be described on an abstract level as the execution of a trajectory. This can be the trajectory towards a goal position like in joint object manipulation or the following of the trajectory like in dancing or motor-rehabilitation. Thus, when two partners collaborate in a haptic task, they have to find a common trajectory for this interaction point or jointly handled object. This implies that the shared action plan towards a task goal in haptic collaboration can be based on the negotiation of trajectories between partners, see also [24] for this consideration. The partners can exchange forces to push or pull in different directions and, by doing so, influence the partner and the shared trajectory. In dependence of the agreement between the partners on the shared trajectory these forces may vary, reflecting the different intentions of the partners.

Based on these consideration we decided in favor of a jointly executed tracking task (see e.g. [31, 59, 60]), where the track represents the desired object trajectory, which is a latent construct in real life applications. The tracking task paradigm offers the advantage that it is non-challenging when it is operated by a single individual as already pointed out by [61]. Hence, when participants are asked to execute a tracking task collaboratively, it is ensured that enough higher cognitive resources are still available to focus on the collaboration with the partner.

For the experimental setup we used a virtual pursuit tracking task and asked participants to move a virtual mass visually represented by a cursor along a given reference path (see Fig. 5.1). As introduced in more detail in the next section, two different conditions, one with and one without haptic feedback from the partner, were defined. This section introduces the aspects of the experimental setup that both conditions have in common.

The reference path was designed as a random sequence of a small number of primitives (triangles, curves, straight lines, jumps). It was visualized as a white line on two black screens (both showing the same scene) and participants were asked to follow this path as accurately as possible with a red cursor representing a virtual mass as the path was scrolling down the screen with a constant velocity of 15 mm/s. One trial took 161 s. Each trial consists of three identical sub-trials and a test curve at the beginning. The horizontal position of the red ball renders the position of the haptic interfaces the participants used to interact with each other.

The cursor position could be manipulated via linear haptic interfaces, one for each partner. These interfaces had one degree of freedom and allowed movements along the x-axis. Each interface was equipped with force sensors, hand knobs and

Fig. 5.1 Experimental setup consisting of two linear haptic interfaces (linked by the virtual mass) and two screens with the graphical representation of the tracking path

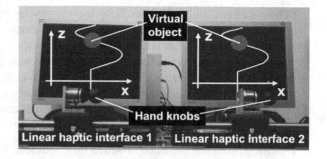

linear actuators. Their control was implemented in Matlab/Simulink and executed on the Linux Real Time Application Interface (RTAI). The graphical representation of the path ran on another computer and communication was realized by a UDP connection in a local area network.

The control of the haptic interfaces was designed to model a jointly carried virtual object. The virtual object was defined to be a mass only and its dynamics was given by:

$$f_{sum}(t) = f_1(t) + f_2(t) = m\ddot{x}_{vo}(t) \tag{5.1}$$

where f_{sum} is the sum of the forces applied by the participants, m is the virtual mass and $\ddot{x}_{vo}$ is the acceleration of the virtual object and, hence, of the linear haptic interfaces. This setup allowed not only the measurement of the resulting force $f_{sum} = f_1 + f_2$ but also of the forces f_1 and f_2 applied by the individual participants.

In order to gain knowledge about the effect of *haptic* collaboration on the dominance distribution in a dyadic tracking task, a condition with haptic feedback from the partner was compared to one without:

(1) *Vision-haptic condition (VH)*: The partners did not only get identical visual feedback of the tracking scenario, but they were also connected via the haptic channel. Beside the mass of the virtual object ($m = 20$ kg) they also felt the forces applied to the object by their partner. This was achieved by introducing a virtual rigid connection between the interacting partners. Thus, $x_{vo}(t) = x_1(t) = x_2(t)$ and the cursor position was determined by

$$x_{vo}(t) = f_{sum}(t) * g_{vo}(t). \tag{5.2}$$

(2) *Vision control condition (V)*, also called two-partners, without-haptic-feedback condition: Again, identical visual feedback was presented to both partners. The mass ($m = 20$ kg) of the cursor was divided into two parts, such that each partner had to carry 10 kg, which presented an equal sharing of workload. The participants felt only the weight of the mass but not the forces applied by their partner. The cursor position was defined as the mean of the two individual device positions

$$x_{vo}(t) = \frac{x_1(t) + x_2(t)}{2}. \tag{5.3}$$

The partners could only infer from inconsistencies between their own movements and the cursor movements on what their partner was doing. The equally splitting of the mass between partners was arbitrary as other definitions of individually presented haptic feedback of the object were possible. Studying dominance as a dyadic construct, we wanted to focus on the interactive component in haptic collaboration and therefore chose this control condition. In contrast to a condition without haptic feedback at all, it was not confounding general effects of this modality with communication specific aspects.

We randomized the sequence in which the conditions were presented to the participants. For a further standardization of the test situation we undertook the following arrangements: a wall was placed between the two participants so they did not gain visual information about the movements of their partner; participants used their right hand to perform the task (all of the participants are right-handed); participants were not allowed to talk to each other during the experiment; white noise was played on the headphones worn by participants, so the noise of the moving haptic interfaces would not distract; the position (left or right seat) was randomized with the order of experimental condition.

Furthermore, participants were instructed to keep the cursor on the path. They were informed about each condition beforehand. In addition, they knew that the first curve of the tracking path was for practice and would be excluded from the analysis.

5.3.3 Measures

As a next step in planing the experiment, ways had to be found to measure dominance behavior. We used the dominance measure proposed by [21], which is force-based and requires to distinguish between different force components.

Given f_1 and f_2, the forces applied by each of the interaction partners, two different force components can be distinguished, the external force f^E and the interactive force f^I

$$f = f^E + f^I. \tag{5.4}$$

Interactive forces occur if the two individuals do not apply forces in the same direction, but push against or pull away from each other. Thus, interactive forces are only defined if the two partners apply forces in opposite directions. In contrast to external forces they do not cause any motion of the object. For a detailed description of these force components and their calculation please refer to [56].

Based on these force components, the individual dominance of partner 1 over 2 (PD_{12}) is defined as

$$PD_{12,i} = \frac{f^E_{1,i}}{f_{1,i} + f_{2,i}} \tag{5.5}$$

where i is the corresponding time step. The same also holds for $PD_{21,i}$. The measure comes with several attributes which are common to most dominance measures in literature: $PD_{12} \in [0, 1]$ and $PD_{12} + PD_{21} = 1$. Hence, a partner is absolutely dominant with a value of one, and absolutely non-dominant with a value of zero. A value of 0.5 means that both partners equally share the workload required to accelerate the object. The individual dominance measure is independent of the direction of the individual forces f_1 and f_2. In contrast to the dominance measure defined in [29], it can be calculated for each time step.

5.3.4 Analysis

As we pointed out in the first part of this chapter, we cannot use standard inference statistic procedures to investigate individual behavior in an interacting dyad. Therefore, we will relate to a model used in social psychology, which overcomes this problem and additionally directly addresses the analysis of behavior constancy across different partners: the social relations model (SRM) introduced in [42, 44, 47, 62]. The method is related to multi-level linear regression, also known as hierarchical linear modeling (HLM), as it introduces random coefficients to the regression model [63, 64]. Thus, SRM explicitly models interdependence between two partners. As the method is not derived by the author, it is stated here for the sake of completeness, but is only summarized in brevity for the given data set. The analysis is based on the mean individual dominance behavior $\overline{PD}$ in the three sub-trials ($j \in \{1, 2, 3\}$). The amount to which partner 1 dominates partner 2 on average per sub-trial and vice versa can be expressed as follows for a given condition:

$$\overline{PD}_{12,j} = \mu + \beta_1 + \gamma_2 + \delta_{12} + \epsilon_{12,j}, \tag{5.6}$$

$$\overline{PD}_{21,j} = \mu + \beta_2 + \gamma_1 + \delta_{21} + \epsilon_{21,j}. \tag{5.7}$$

The parameter μ reflects a fixed effect, namely the mean individual physical dominance measure in a given group. This parameter is of no relevance as the SRM investigates variances. In the following, the first equation referring to partner 1 is described. Partner 2 can be analyzed accordingly. The first random effect is β_1 which presents the actor effect, i.e. the *general tendency of partner 1 to dominate others*, across the j sub-trials and across the different partners in the group. The random effect γ_2 describes the *general tendency of partner 2 to be dominated by others* across the j sub-trials and across the different partners in the group. The third random effect is δ_{12} reflecting the *unique dominance constellation within a specific dyad*, here partner 1 and partner 2 across the sub-trial. The last component $\epsilon_{12,j}$ is the variance in the dominance measure in a given sub-trial j, which cannot be explained by the other components (*error term*). In the given data set there are three different partners per participant. It is important that the SRM is not directly interested in the size of the effect of these components, as there is no causal effect due to specific predictors involved here, as it would be in ordinary regression approaches. Instead, the variance in these effects is the focus of the model. To give an example, the actor variance can be interpreted as an "estimate of the overall amount of variation in dyadic scores that is potentially explainable by characteristics of the individuals who generated them" [47]. Thus, a large variance in the actor effect actually means that changes in the dominance measure are due to characteristics of the actor in contrast to interactive behavior towards the partner.

The goal of the social relations model is to examine the variance of the three random effects (σ_β, σ_γ, σ_δ). Hence, the variance found in all dominance measures in our data set can be partitioned into the three sources (see explanation above), assuming an additive, linear relationship. Furthermore, the SRM distinguishes two types of reciprocity:

(a) *Actor-partner reciprocity* or generalized reciprocity (covariance of β_1 and γ_1 has no meaning when analyzing dominance. This is due to the complementarity of dominance ($\overline{PD}_{12} = (1 - \overline{PD}_{21})$): Negative generalized reciprocity ($\sigma_{\beta,\gamma}$) implies that persons who dominate others are not dominated by others. Thus, the parameters β_1 and γ_1 correlate with $r = -1$ by definition.

(b) *Dyadic-reciprocity* (covariance of δ_{12} and δ_{21}: $\sigma_{\delta,\delta'}$) reflects the unique association between the dominance value of partner 1 and partner 2. This reciprocity provides information on the consistency of the dominance differences in the behavior of two partners across the three sub-trials.

The analysis of the social relations model is conducted using the whole round robin data set. The five variance/covariance parameters of the model are identified with the SOREMO program [65]. All inference statistical results in the next section will be reported on a significance level of 5%.

5.3.5 Results

The results from the SRM-analysis on variances in the mean individual dominance level ($\overline{PD}_{12}$, $\overline{PD}_{21}$) are reported in Table 5.1. The variances of the actor and partner effects (σ_β, σ_γ) are significantly different from zero in both conditions. The average individual dominance behavior, which is consistent across partners, explains 49.0% of the overall variance in the dominance measure in the *V* condition and 64.3% in the *VH* condition. The variances of actor and partner effects are considered together because they both relate to person-dependent behavior, which is not influenced by the interaction itself. The third variance component, the relationship σ_δ, determines the dyad-specific behavior: 32.8% in *V* and 24.7% in *VH*. However, the variation in this effect is not significant in the former condition and, thus, has to be interpreted with care. The higher amount of variance in actor and partner effects compared to the relationship effect implies that the average dominance behavior per interaction is rather person-dependent than due to the interaction with a specific partner.

The actor-partner reciprocity $\sigma_{\beta,\gamma}$ is -1.000 in both conditions which was expected due to the design of the measure in line with the dominance complementarity.

Table 5.1 Model estimates for *relative* variance partitioning (expressed in percentages) for actor, partner and relationship effects and the error term. If the amount of explained variance in $\overline{PD}_{12}$ is significantly different from zero, can be inferred from the p-values given below; significance on a 0.05 level is marked with *

Condition		Actor σ_β	Partner σ_γ	Relationship σ_δ	Error
V	estimates	0.225	0.265	0.328	0.183
	p	0.038*	0.041*	0.051	
VH	estimates	0.353	0.290	0.247	0.110
	p	0.011*	0.022*	0.037*	

The dyadic reciprocity $\sigma_{\delta,\delta'}$ states that the average dominance behavior between partners in the three sub-trials varies in both conditions (V: $\sigma_{\delta,\delta'} = -0.375$; VH: $\sigma_{\delta,\delta'} = -0.523$). Otherwise, a correlation of -1.000 would have been found here as well. However, due to the higher correlation in VH, it can be concluded that with haptic feedback of the partner the mean dominance behavior is more stable across subtrials.

5.3.6 Discussion

The presented analysis allows statements on the consistency of average dominance behavior with a social relations model analysis: A high amount of the variability in average individual dominance behavior is consistent across partners (V: 49.0%, VH: 64.3%). We propose that this result may be explained by a character-trait of participants responsible for the preferred dominance behavior in general, i.e. domineeringness. Hence, an artificial partner can also display a relatively consistent dominance behavior tendency, i.e. take over a certain dominance role (rather dominant or rather non-dominant) without loosing human-like behavioral patterns. However, it is shown that 32.8% of the variance in mean individual dominance behavior in the V condition and 24.7% in the condition with mutual haptic feedback can be explained by interaction between specific partners. We propose that this is the amount the artificial partner has to adapt to the human partner in order to create a natural feeling of interaction in the given task. So far, the adaptability of dominance behavior is based on mean values per interaction. However, the tendency already suggests that control architectures for artificial partners should address these two dominance components (the personal and the interactive component) separately. The fact that the dyadic reciprocity correlation has only small to medium size indicates that the mean dominance difference between partners varies between sub-trials. In the haptic feedback condition the dyadic reciprocity is higher, leading to the conclusion that haptic feedback of the partner provides more stability, and hence, predictability of the average human dominance behavior across time. Therefore, the findings of the consistency analysis support the recommendation to provide mutual haptic feedback: the human dominance behavior is more stable across partner in this case, which enables an easier prediction within collaborative processes.

The analysis presented here is based on an abstract experiment; especially, as the task involved only one-dimensional movements and a track which is not seen in real life scenarios. Thus, the generalizability of the findings to similar experiments or actual applications (external validity) has to be shown by future work. This drawback of the chosen design was accepted in favor of the possibility to investigate the causal relationship between the provided feedback from the partner and the resulting consistency in dominance behavior. Another restriction of the experimental results is the fact that participants were not allowed to speak to each other. A communication channel, which is present in real life and can be additionally provided in virtual environments. The focus on haptic collaboration disregarding verbal communication as done here, eliminates verbal information as potential bias in the collaboration sequence. Again, future experiments have to investigate how the role of the haptic feedback in collaboration is influenced by additional verbal communication. Furthermore, it may be of interest to investigate dominance behavior for a control condition where no haptic feedback at all is given (as e.g. used in [31, 32]), which is a likely scenario in VR.

Future work should address the distribution of dominance in multi-dimensional environments and more complex manipulation tasks. Dominance might be distributed differently in different dimensions (compare e.g. [66, 67] for different responsibilities on dimensions of workspace in human-robot collaboration). Based on the experimental approach presented above, a baseline for these future experiments is given together with an experimental setup and measurements which allow to consider additional or alternative factors and variations.

5.4 Example 2—Late Design Stage: Evaluation of Artificial Partner Implementations

While the last subsection introduced an experiment focusing on knowledge generation and investigated underlying principles of haptic collaboration, the current subsection aims at evaluating first implementations of artificial collaboration partners. Comparing different partner implementations allows to identify their shortcomings and the importance of certain model components. Being an evaluation experiment, this study can be seen as part of later stages in the design process, and therefore, relates to the second approach we introduced in Sect. 5.2.1. This subsection summarizes our work published in [68] by highlighting characteristic challenges of haptic collaboration research.

5.4.1 Research Interest

Today, the design as well as the evaluation of artificial partner implementations focuses mainly on performance, because artificial partners are commonly introduced to assist the human in task execution. In [69], however, it is argued that humans "will be more at ease collaborating with human-like" than with machine-like robots, because they "may be perceived as more predictable" and "human-like characteristics are likely to engender a more human mental model of the robot" when estimating its capabilities. Hence, in the present study, our research interest was not only in evaluating task performance, but also in evaluating human-likeness of the artificial haptic collaboration partner perceived by the human partner. We approached this interest via an advanced "haptic Turing test" that compares artificial partners to a real human.

Most of the studies addressing human-likeness of artificial partners relate to non-haptic interaction. Human-likeness of physical/haptic collaboration partners is approached only in recent studies [29, 70–72]. The effect of providing visual and haptic feedback from a virtual partner on the plausibility of social interaction was evaluated [71]. Plausibility (referring to the perception that an event/object in the virtual world is actually occurring/existing although it is known that it is only computer mediated [73]) was judged on the basis of a 7-point rating scale while performing handshakes with a virtual partner [71]. Similarly, the "Turing-like" evaluation presented by [72] deals with the quality of social interaction in a virtual handshake scenario. They conducted a forced-choice test where the participants had to decide which of two presented models was more human-like. They analyzed their data by fitting a psychophysical function and deriving a human-likeness measure from it. In their experiments, they addressed robot hand-shakes only, no real human condition was introduced as reference.

A "haptic Turing test" to evaluate a feedforward force model in a one degree of freedom rotational pointing task was conducted by [29]. Only discrete yes/no answers were given; the degree of human-likeness on a continuous scale was not addressed. Although not supported by a statistical analysis, they reported that subjectively almost all of the participants "thought they were working with a person," but task performance was worse than in interaction with a real human partner. This approach turns out to be not suited to evaluate more than one artificial partner and to compare different partners to each other in order to decide for the best implementation in a given scenario. Results further point out the necessity of analyzing the relation between subjectively perceived human-likeness and task performance in a more systematic way.

In order to contribute to the design-evaluation cycle of haptic collaboration partners, a measure of human-likeness is needed to provide (a) the degree of human-likeness, and (b) allow a comparison of different partner implementations. This leads to the following two research questions:

How can we measure human-likeness of a artificial collaboration partner and how do the different measures relate to each other?
Which partner implementation is perceived most human-like? What model components are required to achieve high human-likeness?

Fig. 5.2 Difference between
Experiment 1 and 2: Path
with prediction (*left*) and
without prediction (*right*)

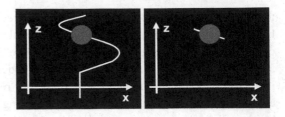

5.4.2 Experimental Approach

In order to investigate these research questions the same experimental setup as adopted in the previous example was used. Only a few modifications were made to fit the setup to the new requirements in line with the new research interest: Since our partner models did not allow for prediction, only the current segment of the tracking path was shown hiding future path segments, see Fig. 5.2. The control condition without haptic feedback was removed as we aimed at comparing existing partner models that provide haptic feedback. In the next subsection the investigated partner models are introduced.

Partner Models and Reference Stimuli

Existing models of haptic collaboration partners are typically based either on feedback or feedforward structures.

Feedforward models are commonly realized by a replay of a pre-recorded motion or force profile. In the context of haptic interaction, one example is the recording of user forces during task execution, which are then replayed by the artificial partner. Feedforward models allow only for unidirectional signal exchange. The pattern replayed does not react or adapt to the behavior of the partner. Reactive behavior is, however, a key feature of haptic collaboration. Hence, if the perception of human-likeness is influenced by adaptive behavior, a model based on pure replay can lead to poor results in a human-likeness evaluation.

To develop haptic collaboration partners that can react to their (human) partner, *feedback* structures have been introduced [23, 24, 74–80]. However, none of these models explicitly takes human haptic collaboration strategies into account.

In this study the following two existing partner models were investigated:

(1) *Partner implementing a feedforward model*: A feedforward structure replaying forces that were recorded while executing the task alone.
(2) *Partner implementing a feedback model*: A feedback structure implementing a McRuer model [74] that aims at minimizing the position tracking error *e*. For the object being a mass, this quasi-linear control model is given by

$$G_h(s) = \frac{F_h(s)}{E(s)} = \underbrace{\frac{e^{-\tau s}}{(1 + T_p s)}}_{\text{perception–action loop}} \left[K(1 + T_z s) \right] \qquad (5.8)$$

where τ is the time-delay caused by the human perception–action loop, T_p is the lag due to the limited bandwidth of the human motor system, and K and T_z are the parameters of the actual human control, cf. [68, 74] for details, or [81] for explanatory comments. The parameters were identified experimentally in our previous work [80]: $K = 18.88$, $T_p = 4.74$, $T_z = 0.12$, and $\tau = 0.12$.

To obtain a scale of human-likeness, two additional experimental conditions were realized to define the upper and lower end of the scale: interaction with a real human partner and interaction with a random signal.

(3) *Human partner*: Interaction with a real human partner defined the upper end of our scale of human-likeness. One trained confederate interacted with all participants. This way, a standardization of behavior was achieved. This procedure was also used by [31, 82–84] and [29].

(4) *Random partner*: A robot interaction partner applying random forces defined the lower end of our human-likeness scale. This unidirectional signal was neither related to task execution nor to human behavior.

Hence, in total four "partners" were investigated. This allowed a comparisons of the two artificial collaboration partners and in addition the identification of their gap to a real human partner. Further, the importance of feedback structures in comparison to feedforward models could be examined. For more details on the implementation of the artificial collaboration partners and reference stimuli please refer to [68].

5.4.3 Measures and Analysis

The human-likeness of a perceived behavior is a latent psychological construct and, hence, not directly measurable. To approach this challenge, we adopted the following three measures.

(1) *Predefined scale*: If the subjective perception of the partner is measured using rating scales (Likert scales), the participants interacted with a partner without knowing its nature (robot or human) and rated its human-likeness afterwards on a given scale. By adding reference stimuli representing the extremes of the human-likeness scale we added a comparative component to the rating-scale procedure and participants were familiar with the range of behavior they had to judge. Through this approach, the degree of human-likeness of different robot partners could be determined and compared to each other. We chose a 5-point human-likeness scale: from 1 = highly non-human/randomly-acting partner to 5 = highly human-like/human partner. In the beginning the participants were presented the two baseline conditions, *human* and *random*.

(2) *Pairwise comparison*: In pairwise comparisons, participants always judged two conditions relative to each other, e.g. "Which condition is more human-like, A or B?" Based on the answers of the participants, an interval scale could be derived. This type of human-likeness evaluation was not based on a predefined

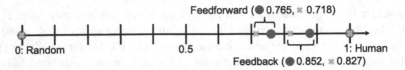

Fig. 5.3 (Color online) Mean human-likeness ratings standardized by linear transformation to the interval [0,1] (*red circle*: subjective ratings; *green cross*: results obtained by pairwise comparisons)

scale, but on a scale generated after the experiment. It was expected that it describes the latent construct of human-likeness better than the scale used in the subjective rating. For the derivation of the interval scale we decided for Thurstone's law of comparative judgment, case 5 [85, 86] out of several methods introduced in the literature.

(3) *Task performance*: Based on the assumption that collaboration with a more human-like partner facilitates performance, task performance was measured as a potential objective measure of this construct. The choice of a task performance measure is usually highly dependent on the particular task. In the present context, dealing mainly with joint object manipulations, typical task performance measures are, e.g., task completion time, time-on-target, and task error [81]. In particular, we used the deviation (root mean square error) of the actual trajectory from the desired reference as task performance measure.

The literature on human-likeness suggests that there is a correlation between human-likeness and task performance, e.g. [87]. If so, task performance could be used as an objective measure of human-likeness. However, preliminary results presented in [29] indicate that task performance and human-likeness are not correlated in haptic collaboration. To take a first important step in finding an answer to this still open research question, we analyzed statistically whether, in our experiment, there is a correlation between human-likeness and task performance.

5.4.4 Results

Subjective Human-Likeness

The human partner was rated most human-like and the random partner was clearly rated worse than any other partner. Furthermore, in both methods the artificial haptic interaction partners were rated highly human-like and the feedforward model was rated slightly less human-like than the feedback model, see Fig. 5.3.

Our evaluation experiment led to different results on *perceived human-likeness* than those presented in [29], where only discrete yes/no answers on human-likeness could be given in a "haptic Turing test." There, a feedforward replay model was perceived to be human-like. We suppose that applying this binary "haptic Turing test" to our experiment would have led to a similar result because participants rated the feedforward condition highly human-like. Hence, if confronted with only two rating options, participants may have decided for "human-like" rather than against it. In our experiment, where a more differentiated rating was possible, it is revealed that there is room for improvement of feedforward models.

Performance

Relating to performance, we could replicate the results found in [29]: In both studies the performance was better with a real human partner. Furthermore, we could show that the virtual partner based on the feedback model gains the same performance results as a real human partner. We explain these findings with the feedback structure's capability to react to errors of the collaboration partner and, hence, support task execution by corrective actions.

Correlation of Subjectively Perceived Human-Likeness and Task Performance

The experimental results of the two subjective human-likeness measures (the predefined scale and the scale derived from pairwise comparisons) showed a high level of agreement in the human-likeness ratings obtained for the experimental conditions. The correlation between subjective human-likeness and task performance revealed that in our experiment only 44% of the variance of one measure can be described by the other.

5.4.5 Discussion

By introducing a real human as a reference stimulus, thus conducting a "Turing test", the gap between human partners and the artificial partners can be investigated. Hence, the potential for future improvements in relation to those artificial partners can be identified. We addressed the differences between a human and two artificial partners with three different (potential) measures of human-likeness. Results reveal, that performance as a possible objective measure is correlated partly with the subjective scales of this latent construct. Hence, we conclude that for a profound evaluation both, performance and subjective measures, should be recorded until this relationship is clarified. The two subjectives measures, however, led to equivalent results.

We consider our study to be a first approach to identifying characteristics of a haptic collaboration partner. On the restricted basis of the current experiment, we conclude that a key feature of advanced haptic partners supporting high task performance is their feedback characteristic. To additionally close the gap in human-likeness to a real human collaboration partner we further suggest a combination of feedforward and feedback models. Such a combination is also motivated by [81], who suggests that human actions are generated by a combination of feedforward and feedback control. The methodology presented in this paper serves as an instrument to investigate the generalizability of our conclusions and to extend the evaluation to further models of haptic collaboration partners.

5.5 Conclusion

This chapter gave an overview on the use of psychological experiments in haptic collaboration research. After describing our view on haptic collaboration and experiments in the design process of artificial collaboration partners, we provided a general overview on challenges related to experiments in this field of research. One main statement in this overview is that psychological experiments can enhance the *identification of requirements* for artificial collaboration partners (earlier stages in design process) and additionally can be employed to *evaluate* existing artificial partners (later stages in design process).

Relating to this theoretical background, we summarized two studies, one for each application area of psychological experiments in the development of haptic collaboration partners. The first experiment strived to *identify requirements* of artificial partners in relation to the consistency of individual dominance behavior across different partners. By investigating two collaborating human partners, we could learn about the required adaptability of artificial partners in terms of dominance. Herein, we contrasted the dominance behavior in a condition with mutual haptic feedback to one where this feedback was only presented from the object, allowing us to investigate differences due to the haptic communication channel. Greater parts in variance of dominance behavior (measuring the mean dominance values per interaction sequence) can be explained by either a partner-invariant source (interpreted as a personal trait, explained the larger part of variance) and a behavior, which changes in dependence of the interaction partner (adaptive component, smaller part of variance). Thus, we suggest that the behavior of artificial partners should address both components as well. Based on the results design guidelines for the dominance behavior of artificial partners were outlined.

The second study we summarized is an example of the usage of psychological experiments in the context of *evaluation* of artificial partners. We used the same experimental setup as before to investigate the human-likeness of two different models of haptic collaboration partners. In order to understand the human-likeness construct more reliably, we employed two different subjective measures of human-likeness and task performance as potential objective measure. Further, we compared the two partners against a random signal and a real human. Thus, the evaluation experiment resembles an advanced Turing test. The two subjective measures led to comparable results in favor of the real human. However, the relation between subjectively perceived human-likeness and task performance, which showed equivalent outcomes for the real human and the feedback-based model of the artificial partner, needs to be investigated further. The results of our study provide hints in favor of a artificial partner containing feedforward as well as feedback structures. Thus, hints on new requirements for interactive partners for jointly executed object manipulation tasks are identified. Based on this new insights, it is possible to conduct future experiments in earlier stages of the design process to find out more about human behavior in terms of these two model components.

Acknowledgements This work was partly supported by the ImmerSence project within the 6th Framework Programme of the European Union, FET—Presence Initiative, contract number IST-2006-027141, see also www.immersence.info.

References

1. Hoffman, G., Breazeal, C.: Robots that work in collaboration with people. In: AAAI Fall Symposium on the Intersection of Cognitive Science and Robotics (2004)
2. Bratman, M.: Shared cooperative activity. Philos. Rev. **101**(2), 327–341 (1992)
3. Grosz, B., Sidner, C.: Plans for discourse. In: Intentions in Communications, pp. 417–444. MIT Press, Cambridge (1990)
4. Kanno, T., Nakata, K., Furuta, K.: A method for team intention inference. Int. J. Hum.-Comput. Stud. **58**, 394–413 (2003)
5. Johannsen, G., Averbukh, E.A.: Human performance models in control. In: Proceedings of International Conference on Systems, Man and Cybernetics, 1993. Systems Engineering in the Service of Humans (1993)
6. Grosz, B.: Collaborative systems. AI Mag. **17**, 67–85 (1996)
7. Tomasello, M., Carpenter, M., Call, J., Behne, T., Moll, H.: Understanding and sharing intentions: The origins of cultural cognition. Behav. Brain Sci. **28**, 675–735 (2005)
8. Meulenbroek, R., Bosga, J., Hulstijn, M., Miedl, S.: Joint-action coordination in transferring objects. Exp. Brain Res. **180**(2), 333–343 (2007)
9. Schubö, A., Vesper, C., Wiesbeck, M., Stork, S.: Movement coordination in applied human-human and human-robot interaction. In: HCI and Usability for Medicine and Health Care, pp. 143–154. Springer, Berlin (2007)
10. Sebanz, N., Bekkering, H., Knoblich, G.: Joint action—Bodies and minds moving together. Trends Cogn. Sci. **10**(2), 70–76 (2003)
11. Welsh, T.: When $1 + 1 = 1$: The unification of independent actors revealed through joint Simon effects in crossed and uncrossed effector conditions. Hum. Mov. Sci. **28**(6), 726–737 (2009)
12. Nudehi, S., Mukherjee, R., Ghodoussi, M.: A shared-control approach to haptic interface design for minimally invasive telesurgical training. IEEE Trans. Control Syst. Technol. **13**, 588–592 (2005)
13. Esen, H., Sachsenhauser, A., Yano, K., Buss, M.: A multi-user virtual training system concept and objective assessment of trainings. In: The 16th IEEE International Symposium on Robot and Human Interactive Communication (Ro-Man) (2007)
14. Bettini, A., Marayong, P., Lang, S., Okamura, A., Hager, G.: Vision-assisted control for manipulation using virtual fixtures. IEEE Trans. Robot. **20**(6), 953–966 (2004)
15. Kragic, D., Marayong, P., Li, M., Okamura, A., Hager, G.: Human-machine collaborative systems for microsurgical applications. Int. J. Robot. Res. **24**(9), 731–741 (2005)
16. Butler-Bowdom, T.: 50 Psychology Classics: Who We Are, How We Think, What We Do; Insight and Inspiration from 50 Key Books. Nicholas Brealey, London (2006)
17. Shaughnessy, J.: Research Methods in Psychology, 8th edn. Mcgraw-Hill College, Boston (2008)
18. Wundt, W.: Grundzüge der physiologischen Psychologie (Principles of Physiological Psychology). VDM Verlag Dr. Müller, Saarbrücken (1874), Auflage von 2007
19. Butler, J., Holden, K., Lidwell, W.: Universal Principles of Design: 100 Ways to Enhance Usability, Influence Perception, Increase Appeal, Make Better Design Decisions, and Teach Through Design. Rockport, Gloucester (2007)

20. Sharp, H., Rogers, Y., Preece, J.: Interaction Design: Beyond Human-Computer Interaction. Wiley, New York (2007)
21. Rahman, M., Ikeura, R., Mizutani, K.: Cooperation characteristics of two humans in moving an object. Mach. Intell. Robot. Control **2**(4), 43–48 (2002)
22. Reed, K., Peshkin, M., Hartmann, M., Patton, J., Vishton, P., Grabowecky, M.: Haptic cooperation between people, and between people and machines. In: IEEE/RSJ International Conference on Intelligent Robots and Systems (IROS) (2006)
23. Corteville, B., Aertbelien, E., Bruyninckx, H., De Schutter, J., Van Brussel, H.: Human-inspired robot assistant for fast point-to-point movements. In: IEEE International Conference on Robotics and Automation (2007)
24. Evrard, P., Kheddar, A.: Homotopy switching model for dyad haptic interaction in physical collaborative tasks. In: Proc. of the Third Joint Eurohaptics Conference and Symposium on Haptic Interfaces for Virtual Environment and Teleoperator Systems, pp. 45–50 (2009)
25. Wilson, J., Rutherford, A.: Mental models: Theory and application in human factors. Hum. Factors **31**(6), 617–663 (1989)
26. Cooper, A., Reitman, R., Cronin, D.: About Face 3: The Essentials of Interaction Design. Wiley, New York (2007)
27. Galitz: The Essential Guide to User Interface Design: An Introduction to GUI Design Principles and Techniques. Wiley, New York (2007)
28. Wolpert, D., Doya, K., Kawato, M.: A unifying computational framework for motor control and social interaction. Philos. Trans. R. Soc. Lond. B, Biol. Sci. **358**(1431), 593–602 (2003)
29. Reed, K.B., Peshkin, M.: Physical collaboration of human-human and human-robot teams. IEEE Trans. Haptics **1**, 108–120 (2008)
30. Feth, D., Tran, B., Groten, R., Peer, A., Buss, M.: Shared-control paradigms in multi-operator-single-robot teleoperation. In: Cognitive Systems Monographs, pp. 53–62. Springer, Berlin (2009)
31. Basdogan, C., Ho, C.H., Srinivasan, M.: An experimental study on the role of touch in shared virtual environments. ACM Trans. Comput.-Hum. Interact. **7**, 443–460 (2000)
32. Sallnäs, E.L., Rassmus-Gröhn, K., Sjöström, C.: Supporting presence in collaborative environments by haptic force feedback. ACM Trans. Comput.-Hum. Interact. **7**(4), 461–476 (2000)
33. Schauss, T., Groten, R., Peer, A., Buss, M.: Evaluation of a coordinating controller for improved task performance in multi-user teleoperation. In: Haptics: Generating and Perceiving Tangible Sensations. Lecture Notes in Computer Science, vol. 6191, pp. 240–247 (2010)
34. Sallnäs, E.L.: Improved precision in mediated collaborative manipulation of objects by haptic force feedback. In: Proceedings of the First International Workshop on Haptic Human-Computer Interaction (2001)
35. Groten, R., Feth, D., Klatzky, R., Peer, A., Buss, M.: Efficiency analysis in a collaborative task with reciprocal haptic feedback. In: The 2009 IEEE/RSJ International Conference on Intelligent Robots and Systems (2009)
36. Groten, R., Feth, D., Peer, A., Buss, M.: Shared decision making in a collaborative task with reciprocal haptic feedback—an efficiency-analysis. In: IEEE International Conference on Robotics and Automation (2010)
37. Feth, D., Groten, R., Peer, A., Hirche, S., Buss, M.: Performance related energy exchange in haptic human-human interaction in a shared virtual object manipulation task. In: Third Joint EuroHaptics Conference and Symposium on Haptic Interfaces for Virtual Environment and Teleoperator Systems (2009)
38. Pan, P., Lynch, K., Peshkin, M., Colgate, J.E.: Human interaction with passive assistive robots. In: IEEE 9th International Conference on Rehabilitation Robotics (2005)
39. Schroeder, R., Steed, A., Axelsson, A.S., Heldal, I., Abelin, A., Widström, J., Nilsson, A., Slater, M.: Collaborating in networked immersive spaces: as good as being there together? Comput. Graph. **25**(5), 781–788 (2001)
40. Kashy, D., Snyder, D.: Measurement and data analytic issues in couples research. Psychol. Assess. **7**(3), 338–348 (1995)
41. Griffin, D., Gonzalez, R.: Correlational analysis of dyad-level data in the exchangeable case. Psychol. Bull. **118**, 430–439 (1995)

42. Kenny, D.: The design and analysis of social-interaction research. Annu. Rev. Psychol. **47**, 59–86 (1996)
43. Maguire, M.C.: Treating the dyad as the unit of analysis: A primer on three analytic approaches. J. Marriage Fam. **61**(1), 213–223 (1999)
44. Kenny, D., Mohr, C., Levesque, M.: A social relations variance partitioning of dyadic behavior. Psychol. Bull. **127**(1), 128–141 (2001)
45. DeCoster, J.: Using anova to examine data from groups and dyads (2002). http://www.stat-help.com/notes.html
46. Griffin, D., Gonzalez, R.: Models of dyadic social interaction. Philos. Trans. R. Soc. Lond. B, Biol. Sci. **358**(1431), 573–581 (2003)
47. Kenny, D., Kashy, D., Cook, W.: Dyadic Data Analysis. The Guilford Press, New York (2006)
48. Field, A., Hole, G.: How to Design and Report Experiments. Sage, London (2002)
49. Tabachnick, B., Fidell, L.: Experimental Designs Using ANOVA. Brooks/Cole, Pacific Grove (2006)
50. Tabachnick, B., Fidell, L.: Using Multivariate Statistics. Pearson Education, Upper Saddle River (2006)
51. Howell, D.: Fundamental Statistics for the Behavioral Sciences. Wadsworth, Belmont (2007)
52. Tullis, T., Albert, B.: Measuring the User Experience. Morgan Kaufman, San Mateo (2008)
53. Rubin, J., Chisnell, D.: Handbook of Usability Zesting: How to Plan, Design, and Conduct Effective Tests, 2nd edn. Wiley, New York (2008)
54. Field, A.: Discovering Statistics Using SPSS. Sage, London (2009)
55. Groten, R., Peer, A., Buss, M.: Interpretation of results in experimental haptic interaction research. In: Workshop on Haptic Human-Robot Interaction. IEEE/RSJ International Conference on Intelligent Robots and Systems (IROS) (2009)
56. Groten, R., Feth, D., Goshy, H., Peer, A., Kenny, D., Buss, M.: Experimental analysis of dominance in haptic collaboration. In: The 18th International Symposium on Robot and Human Interactive Communication (2009)
57. Rogers-Millar, L., Millar, F.: Domineeringness and dominance: A transactional view. Hum. Commun. Res. **5**, 238–246 (1979)
58. Burgoon, J., Johnson, M., Koch, P.: The nature and measurement of interpersonal dominance. Commun. Monogr. **65**, 308–335 (1998)
59. Glynn, S., Henning, R.: Can teams outperform individuals in a simulated dynamic control task. In: Proceedings of the Human Factors and Ergonomics Society Annual Meeting (2000)
60. Glynn, S., Fekieta, R., Henning, R.A.: Use of force-feedback joysticks to promote teamwork in virtual teleoperation. In: Proceedings of the Human Factors and Ergonomics Society Annual Meeting (2001)
61. Rasmussen, J.: Skills, rules, and knowledge; signals, signs, and symbols, and other distinctions in human performance models. IEEE Trans. Syst. Man Cybern. **13**, 257–266 (1983)
62. Bond, C., Lashley, B.: Round-robin analysis of social interaction: Exact and estimated standard errors. Psychometrika **61**(2), 303–311 (1996)
63. Snijeders, T., Kenny, D.: The social relations model for family data: A multilevel approach. Pers. Relatsh. **6**(4), 471–486 (2005)
64. Gelman, A., Hill, J. (eds.): Data Analysis Using Regression and Multilevel/Hierarchical Models. Cambridge University Press, Cambridge (2008)
65. Kenny, D.A.: Soremo [computer program] (1994). http://davidakenny.net/srm/srmp.htm
66. Wojtara, T., Uchihara, M., Murayama, H., Shimoda, S., Sakai, S., Fujimoto, H., Kimura, H.: Human-robot cooperation in precise positioning of a flat object. In: Proceedings of the 17th World Congress The International Federation of Automatic Control (2008)
67. Wojtara, T., Uchihara, M., Murayama, H., Shimodaa, S., Sakaic, S., Fujimotoc, H., Kimura, H.: Human-robot collaboration in precise positioning of a three-dimensional object. Automatica **45**(2), 333–342 (2009)
68. Feth, D., Groten, G., Peer, A., Buss, M.: Haptic human-robot collaboration: Comparison of robot partner implementations in terms of human-likeness and task performance. Presence **20**(2), 173–189 (2011)

90   R. Groten et al.

69. Hinds, P., Roberts, T., Jones, H.: Whose job is it anyway? A study of human-robot interaction in a collaborative task. Hum.-Comput. Interact. **19**(1), 151–181 (2004)
70. Ikeura, R., Inooka, H., Mizutani, K.: Subjective evaluation for maneuverability of a robot cooperating with humans. J. Robot. Mechatron. **14**(5), 514–519 (2002)
71. Wang, Z., Lu, J., Peer, A., Buss, M.: Influence of vision and haptics on plausibility of social interaction in virtual reality scenarios. In: Haptics: Generating and Perceiving Tangible Sensations. Proceedings of International Conference. Part II. EuroHaptics 2010, Amsterdam, 8–10 July 2010. Lecture Notes in Computer Science, pp. 172–177. Springer, Berlin (2010)
72. Karniel, A., Nisky, I., Avraham, A., Peles, B.C., Levy-Tzedek, S.: A Turing-like handshake test for motor intelligence. In: Haptics: Generating and Perceiving Tangible Sensations. Proceedings of EuroHaptics 2010 International Conference. Part I, Amsterdam, 8–10 July 2010. Lecture Notes in Computer Science, vol. 6192, pp. 197–204. Springer, Berlin (2010)
73. Slater, M.: Place illusion and plausibility can lead to realistic behaviour in immersive virtual environments. Philos. Trans. R. Soc. Lond. B, Biol. Sci. **364**(1535), 3549–3557 (2009)
74. McRuer, D., Jex, H.: A review of quasi-linear pilot models. IEEE Trans. Hum. Factors Electron. **HFE-8**(3), 231–249 (1967)
75. Flash, T., Hogan, N.: The coordination of arm movements: An experimentally confirmed mathematical model. J. Neurosci. **5**(7), 1688–1703 (1985)
76. Kazerooni, H., Guo, J.: Human extenders. J. Dyn. Syst. Meas. Control **115**(2B), 281–290 (1993)
77. Rahman, M., Ikeura, R., Mitzutani, K.: Investigation of the impedance characteristic of human arm for development of robots to cooperate with humans. JSME Int. J. Ser. C **45**(2), 510–518 (2002)
78. Duchaine, V., Gosselin, C.: General model of human-robot cooperation using a novel velocity based variable impedance control. In: Proc. of the 2nd Joint EuroHaptics Conference and Symposium on Haptic Interfaces for Virtual Environment and Teleoperator Systems, pp. 446–451 (2007)
79. Duchaine, V., Gosselin, C.: Safe, stable and intuitive control for physical human-robot interaction. In: Proc. of the IEEE International Conference on Robotics and Automation, pp. 3383–3388 (2009)
80. Feth, D., Groten, R., Peer, A., Buss, M.: Control-theoretic model of haptic human-human interaction in a pursuit tracking task. In: Proc. of the 18th IEEE International Symposium on Robot and Human Interactive Communication, pp. 1106–1111 (2009)
81. Jagacinski, R., Flach, J.: Control Theory for Humans—Quantitative Approaches to Modeling Performance. Lawrence Erlbaum Associates, New Jersey (2003)
82. Gentry, S., Wall, S., Oakley, I., Murray-Smith, R.: Got rhythm? Haptic-only lead and follow dancing. In: Proc. of EuroHaptics, pp. 481–488 (2003)
83. Allison, R., Zacher, J., Wang, D., Shu, J.: Effects of network delay on a collaborative motor task with telehaptic and televisual feedback. In: Proc. of the ACM SIGGRAPH International Conference on Virtual Reality Continuum and Its Applications in Industry, pp. 375–381 (2004)
84. Khademian, B., Hashtrudi-Zaad, K.: Performance issues in collaborative haptic training. In: Proc. of the IEEE International Conference on Robotics and Automation, pp. 3257–3262 (2007)
85. Thurstone, L.L.: A law of comparative judgment. Psychol. Rev. **34**(107), 4 (1927)
86. Gescheider, G.: Psychophysics: Method, Theory, and Application. Lawrence Erlbaum Associates, Hillsdale (1985)
87. Feil-Seifer, D., Skinner, K., Mataric, M.: Benchmarks for evaluating socially assistive robotics. Interact. Stud. **8**(17), 423–439 (2007)

Chapter 6
Human-Robot Adaptive Control of Object-Oriented Action

Satoshi Endo, Paul Evrard, Abderrahmane Kheddar, and Alan M. Wing

Abstract This chapter is concerned with how implicit, nonverbal cues support co-ordinated action between two partners. Recently, neuroscientists have started uncovering the brain mechanisms involved in how people make predictions about other people's behavioural goals and intentions through action observation. To date, however, only a small number of studies have addressed how the involvement of a task partner influences the planning and control of one's own purposeful action. Here, we review three studies of cooperative action between human and robot partners that address the nature of predictive and reactive motor control in cooperative action. We conclude with a model which achieves motor coordination by task partners each adjusting their actions on the basis of previous trial outcome.

6.1 Introduction

Social skills are what largely distinguish humans from other animals and they are considered to underpin the development of our civilization in the process of evolution [1, 2]. It has been suggested that altruistic interaction among organisms in kin selection or prosocial reciprocity increases the chance of 'survival' [3]. In particular, cooperation, an act of working towards and achieving a shared goal with other agents, is a useful strategy to accomplish tasks that are otherwise inefficient or impossible to complete by a single person. From a psychological perspective, there is

S. Endo (✉) · A.M. Wing
Behavioural Brain Sciences Centre, University of Birmingham, Edgbaston, Birmingham B15 2TT, UK
e-mail: s.endo@bham.ac.uk

P. Evrard
Laboratoire d'Intégration des Systémes et des Technologies (LIST), Commissariat à l'énergie atomique et aux énergies alternatives (CEA), Fontenay-aux-Roses, France

A. Kheddar
CNRS-AIST Joint Robotics Laboratory (JRL), UMI3218/CRT, Tsukuba, Japan

A. Kheddar
CNRS-UM2 LIRMM, Interactive Digital Human, Montpellier, France

A. Peer, C.D. Giachritsis (eds.), *Immersive Multimodal Interactive Presence*,
Springer Series on Touch and Haptic Systems,
DOI 10.1007/978-1-4471-2754-3_6, © Springer-Verlag London Limited 2012

growing research interest in so-called *joint action*, an umbrella term which covers any form of social interaction between two or more people [4]. A series of studies on joint action has provided compelling evidence that the manner in which the central nervous system (CNS) represents the external world is constantly influenced by the presence of others. In particular, actions performed by others elicit similar neural changes to when the action is performed by oneself [5, 6]. It has been proposed that this alteration of neural state is evidence for emulation of others' actions using one's own motor system, so assisting humans to appreciate action intentions of others and to engage in cooperative or competitive behaviours [7, 8]. While these studies have advanced our understanding of the cognitive architecture of the perceptual-motor systems that support joint action, only a small number of studies have directly addressed how the involvement of a task partner influences the planning and control of one's own purposeful action (e.g., [9–11]). In this chapter, we describe recent developments in our understanding of interpersonal interaction and present new studies that highlight predictive and reactive motor control during cooperative action.

6.2 Feedforward and Feedback Control in Single-Person Movement

Sensory-motor systems are subject to internal errors, including inaccurate target specification due to sensory noise and faulty movement execution due to motor noise, which manifests as fluctuations of movement over repetitions [12]. One approach to overcome such error is through on-line corrections wherein on-going movements are continuously monitored using sensory feedback and any difference between intended and actual movement is corrected whenever an error is detected. A weakness in this feedback control scheme is that the motor response is subject to a significant time delay caused by the relay of the sensory signals from the sensors to the cerebral cortex and the subsequent implementation of an appropriate response. In the case of object-lifting with precision grip (opposed thumb and index finger) by a single actor, somatosensory feedback about object slippage on the digits induces an adjustment of the grasping force to prevent further slippage, which lags at least 80–100 ms behind the onset of the slippage [13]. Given that the spinal reflex takes about 30 ms to trigger a reflexive response [14], the motor correction in response to slip is considered to require supraspinal structures. Another method for overcoming motor error is through learning and adaptation. In feedforward control, the CNS implements action using a model containing information regarding the expression of the motor commands in the environment and their action consequences, information learned from previous interactions in the same or similar environments [15]. An appropriate motor prediction can then be generated by inverting this internal model. Some theories of feedforward control suggest that the CNS generates predictions about the sensory consequences of a given action in order to help make that action become more accurate [16]. Hence, when the same movement is repeated, the CNS learns and adapts the movement to reduce the chance of making the same error by adjusting motor outputs [15]. It is thought that feedforward and feedback control are flexibly integrated depending on the nature of the task and its familiarity. In general, the CNS initially implements a control policy that is more reliant on feedback control and it gradually shifts towards feedforward control as the actor learns the task.

6.3 Sensory-Motor Control in Cooperative Action

In contrast to individual action, cooperative action between two actors involves two independent sets of movements that are coordinated in time and space. Thus, the manner in which two partners intend to perform an action together may not at first be complementary to each other. A verbal exchange could be sufficient to share a general motor strategy for action such as the path to follow in picking up and moving an object together. Nevertheless, cooperative action with another person does not necessarily require conversation; other forms of sensory input can provide sources of information enabling prediction of a partner's action. As noted above, sensory-motor systems are subject to internal errors, including inaccurate target specification due to sensory noise and faulty movement execution due to motor noise [12]. When the motor outcome deviates appreciably from the intended movement, the CNS adapts and refines the output signal to minimise the risk of making the same error again [15]. Such adaptive behaviour can be accounted for by simple linear models [17, 18]. In these models, motor output in a subsequent response is adjusted by some proportion of the observed error from the desired state in the previous movement. In principle, various adaptive models conform to this structure [19–21]. This 'error-based learning' plays a key role in optimising anticipatory movements in individual action [17]. Similarly, evidence indicates that interpersonal motor adaptation may be based on previous performance error during cooperative action. In a study by Schuch and Tipper [22], participants either performed a speeded choice-reaction-time task or observed another person performing the same task. When a participant performed the task and made a mistake, the reaction time in the subsequent trial was slowed, a phenomenon termed as post-error slowing [23]. Interestingly, their results showed that the reaction time was also slowed after observing somebody else making the error. This study demonstrates that observing another person's error influences that observer's subsequent behaviour, strengthening the argument for the existence of common mechanisms for processing one's own errors and the errors of others [24]. In a recent study in our own lab we have more direct evidence of error-based learning in joint action involving cooperative lifting of a rigid bar. We examined cooperative object lifting using a humanoid robot as a task partner to precisely control behaviour of one of the partners so we could understand how a particular movement characteristic of the partner (i.e., robot) influences the action of the human participant. Participants sat at a table and grasped a 6 DoF force transducer attached to one end of a bar; a robot held the other (see Fig. 6.1(a)). Their task was to lift the bar so it remained horizontal throughout the lift. The robot was programmed to produce a vertical movement trajectory based on a minimum-jerk

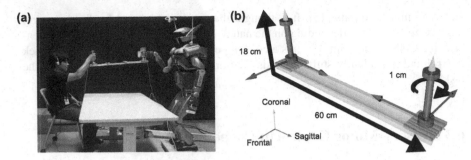

Fig. 6.1 (**a**) Experimental setup. HRP-2 (Kawada Industries, Japan) is a full-sized humanoid robot with 30 degrees of freedom (DoF). For this lifting task, only the 6 DoF of the right arm were used. A participant sat in front of the robot and jointly lifted a bar. (**b**) A schematic illustration of the bar. The bar consisted of two handles with force transducers attached onto their bases. The orientation of the transducers was expressed in ego-centric coordinates with respect to each partner

trajectory [25] with a target movement height and duration of 40 cm over 3 seconds. In this, and the following studies, participants closed their eyes and wore a pair of headphones to prevent any visual or auditory feedback during performance.

We measured the maximum difference in the height of the two handles (i.e., position error), difference in the peak velocities of the partners (i.e., velocity error), and the maximum sagittal force and frontal torque (task-relevant) in order to study whether participants could improve cooperative performance with practice. Figure 6.2 shows changes in these variables over 15 repetitions (averaged over 10 right-handed participants). Performance rapidly improved over successive trials, as shown by the reduction over a couple of trials in the position and velocity errors. The sagittal force reduced over trials, but more slowly over many trials. Thus, the smaller sagittal force is unlikely to be a simple by-product of reduced coordination error. Importantly, the sagittal force exchange between the partners is redundant (i.e., task irrelevant) to the primary goal of the task. Therefore, the reduction of the sagittal force may be linked to minimising redundant interaction in cooperative action [26].

The results so far show evidence for motor adaptation to a task partner, when the robot repetitively executes an identical motion. However, performance of a human partner may vary from trial to trial. For example, various factors can influence one's action including a variety of internal factors such as sensory and motor noises in the CNS [12], exhaustion [27], or individual preference as well as external factors such as the presence of an obstacle in the movement path [28]. Thus, in another experiment, we investigated how people minimise an unpredictable motor error induced by a task partner. In this part of the study, the robot randomly introduced two perturbed trajectories (Retarded and Advanced, Fig. 6.3(a)) in addition to the previously used unperturbed trajectory (Standard). In Retarded trajectories, there was a sudden linear retardation in the velocity profile such that 70 ms after the perturbation onset, the vertical position of the gripper was 1.0 cm lower than at the equivalent time in the Standard trajectory. The Advanced trajectory was analogous to the Retarded trajectory except that this perturbation caused an advance of the time of peak

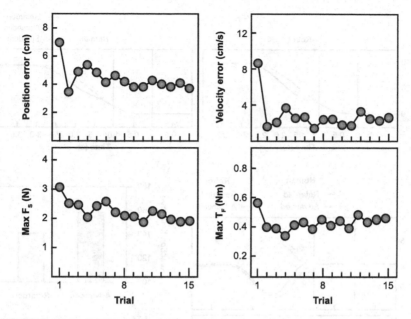

Fig. 6.2 Learning cooperative lifting. The position error, velocity error, maximum sagittal force (F_S), and maximum frontal torque (T_F) of participants when interacting with each trajectory type presented in the different conditions/blocks of trials

velocity. The perturbation trials were randomly inserted between the Standard trials but they never occurred in adjacent trials. There were 15 Advanced and 15 Retarded trials and 90 Standard trials.

Figure 6.3(b) shows the averaged kinematic differences between the participants and the robot. When an unpredictable robot movement was detected during lifting, the participants still managed to follow closely the robot's trajectory after an initial delay of approximately 150 ms (Fig. 6.3(c)). This is noticeably longer than the 80 ms or so taken, according to the literature, by the supraspinal reflex for movement correction via the somatosensory system [13]. Thus, in response to a sudden perturbation, the participants adjusted their on-going action to re-synchronise their own action to their partner's action, though the response was delayed possibly due to complex haptic feedback underlying cooperative object lifting. We recently obtained evidence of such use of haptic feedback in a two-person object lifting study in which we removed partners' torque feedback by adding ball bearings at the grasp points. This significantly disrupted motor coordination and between-trial learning compared to two-person object lifting task with fixed grasp points [29]. In the human-robot cooperative lifting study, the on-line error correction was smoothly introduced and completed at around the deceleration phase of the lifting movement, with the Advanced trials associated with slightly faster recovery. Taken together, the two human-robot cooperative lifting studies demonstrate that the participants performed cooperative action using a form of control that could be flexibly switched

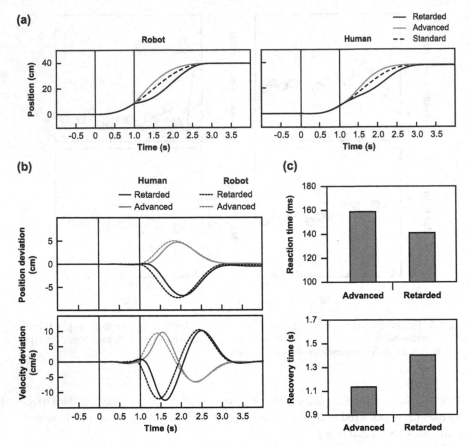

Fig. 6.3 (a) An example of movement trajectories by the robot and participant. The robot's control signal instead of the recorded motion data was depicted. Averaged Standard, Advance, and Retardation trajectories from a single participant are presented. (b) Kinematic divergences of averaged Perturbation trajectories from the Standard trajectory of a single participant. The position and velocity profiles of the Standard trajectory were subtracted from the Perturbation trajectories. (c) The reaction time to, and recovery time from, the Advance and Retardation perturbations

between predictive and reactive modes depending on the reliability of the partner's movement.

6.4 Mutual Error-Based Learning in Cooperation

The above studies demonstrate between and within trial adaptation that people use to cooperate with a robot task partner that did not adapt to the human partner. Thus, in this sense at least, the partnership was asymmetric. In practice, purposeful interpersonal interaction does often involve asymmetric yet complementary actions. Simulation studies have suggested that asymmetric involvement in a task can effectively reduce conflict in cooperative interaction [30, 31]. Furthermore, differential role assignment has been associated with more flexible adaptation to environmental changes [32]. Behaviourally, functional asymmetry has been demonstrated in a situation where two people each apply force to achieve a shared goal. Reed and colleagues [11] used a device that allowed recording of forces created by two partners, which was a rotating table with two handles attached at each end. Using this device, paired participants grasped and moved one handle each to meet the visual target as quickly as possible. Even though the participants were instructed to refrain from verbal communication, they rapidly developed a strategy such that one was more involved in accelerating the device and the other in deceleration. The study by these authors, therefore, indicates that functional specialisation can be observed at an interpersonal level of action coordination. Hence orchestrating two effectors by functionally specialising the effectors in subtasks rather than producing identical movements may be an effective strategy. In motor interaction between humans, the movement coordination of two people is inter-dependent. Thus, the task partners may need to compromise their own individually preferred courses of action and adapt their own motor output to that of their task partner [33–35]. For example, either intentionally or unintentionally, the presence of others influences the timing of our behaviour. Thus, behavioural synchronicity of people, termed entrainment, has been reported in various forms and sizes of interaction such as dyadic walking rhythm [36, 37], body gestures [38] and conversational pattern [39]. In a study by Richardson and colleagues [40], for example, paired participants sat on a rocking chair side-by-side and rocked the chair at a preferred frequency. Without instruction to do so, the participants started synchronising their movements to each other. On a much larger scale, spontaneous interpersonal movement synchronisation was reported in the hand clapping pattern of thousands of members of the audience at a concert hall [41]. Frequently, behavioural interactions of two people have been modelled in terms of a dynamical system. Namely, it has been proposed that two different sets of movements interact and mutually influence each other's state in order to find a stable solution [42, 43] or reduce their differences [20, 44] in temporal motor coordination tasks. These studies provide specific predictions about how people may coordinate their movements with each other. However, the interaction process has not previously been studied in terms of cooperative goal-directed action such that the mutual interaction should relate to optimal performance of a cooperative outcome rather than merely reflect coordination of two sets of movements. In contrast, we recently studied how adaptation by a task partner affects own motor adaptation. In this joint adaptation task, the participant lifted an object with the robot, but now the robot's speed of the lifting was determined by a simple model, which reduced the peak velocity difference between the partners:

$$Z_{Rn+1} = Z_{Rn} + \alpha_R(Z_{Hn} - Z_{Rn}) \tag{6.1}$$

where Z_{Rn} is the peak velocity of the robot at Trial n and α_R is the adaptation rate which reduces the mismatch with a human partner. Over trials the adaptation rate of the robot was either fixed ($\alpha_R = 0.8, 0.5, 0.2$, or 0.0), or varied randomly between 0.0 and 1.0 in 0.1 increments. In order to keep the bar level the human participant

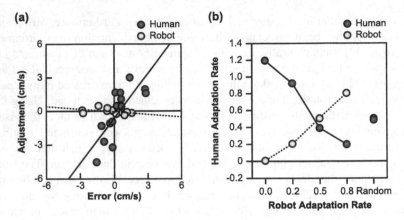

Fig. 6.4 (a) A scatter plot of an error and subsequent adjustment in the peak velocity from a single participant and the robot. The slope coefficient of the least-square-error represented the adaptation rate. (b) The adaptation rates of human participants across different rates set by the robot

would need to match the velocity of the robot. One way to do this would be to adjust the peak velocity;

$$Z_{Hn+1} = Z_{Hn} + (w - \alpha_R)(Z_{Rn} - Z_{Hn}) \qquad (6.2)$$

where w is the net adaptation rate which would be set to unity for the most rapid reduction in error. It was therefore predicted that participants would increase their adaptation rate when that of the robot was small, and vice versa. In the case of the randomly varying rate it was expected the human partner's adaptation rate would equal the average rate of the robot. On the other hand, if the adaptation rate of the robot was not taken into account, it was expected that the adaptation rate of the human partner would be similar across the different correction gains of the robot. The adaptation rate of each participant (α_H) was approximated using the slope coefficient of the linear regression, with interpersonal difference of the peak velocity at a previous trial and the peak velocity adjustment made from this trial to the next trial as variables (Fig. 6.4(a)).

The results show that α_H generally varied inversely with α_R (Fig. 6.4(b)). Simple linear regression analysis confirmed that there was a negative relationship between α_H and α_R such that the increase of α_R led to a decrease of α_H. The analyses provided clear evidence that human participants flexibly modulated their correction gain so that the summation of α_R and α_H (α_{NET}) remained constant at around -1. This study provides clear evidence that people can implicitly and flexibly adjust their adaptation rates with regard to the adaptation rate of their partner in order to optimise the performance at a cooperative level. This modulation of adaptation rate may crucially underlie social interaction, which enables people to respond flexibly to other people's actions in a wide variety of social interactions and thus facilitate goal-directed behaviour in joint actions. In this study, we used the regression between the motor error and subsequent adjustments to estimate adaptation rate. However, this method is susceptible to increasing bias as proportion of the error (adjustment) signal over the sensory-motor noise decreases [45]. However, this problem can be rectified by inserting random perturbations and measuring resulting responses and we recently completed a new study utilising this method [46]. This study also showed complementary gain setting, albeit with lower adaptation gain than seen when cooperating with the robot.

6.5 Determinants of an Adaptation Rate

This joint adaptation study demonstrated that people can effectively accommodate contribution of an adaptive partner in a cooperative task so that the summed adaptation rates of the partners are sustained at a certain level. However, it remains unclear how a dyad delegates the adaptation rates between its respective partners. Recently, Braun and colleagues [47] have described cooperative behaviour in terms of efficient decision making. In their study, two participants jointly moved a handle to a target, but the force required to move the handle in space varied. When they chose a "cooperative" path, the participants both received resisting spring force of 3 N/m towards the start position as they moved the handle to its target. When they both chose a "defective" path, they experienced a higher spring force of 7 N/m. When one person selected the "cooperative" path but the other one did not, the former received 10 N/m spring force and the latter experienced no spring resistance (0 N/m). Under such circumstances, the participants chose the "defective" path whereby their partner would never be at an advantage. When they performed the same task bimanually, in contrast, the participants actively achieved the cooperative solution. Thus it seems that people are estimating costs of relying on their partner when they make a decision about how they should perform the task. Although this study showed a tendency for people to optimise their individual action over cooperation, there is evidence suggesting that people do not simply discount the performance of others, but rather track the probability of their partner's reliability in making a decision. In a study by Behrens and colleagues [19], participants performed a decision-making task wherein a correct response could be predicted by the pattern of previous correct responses as well as by a clue provided by a confederate. The reliabilities of the reward-based learning and confederate advice were independently varied to study how the participants' decision-making was influenced by the statistical properties of the reward prediction error and confederate prediction error. The results revealed that the participants could learn both reward and social value in a similar fashion but independently. This study provides strong evidence that the optimal response in decision-making is in consideration of both probability of the correct response and the probability of the correct advice by the confederate. Thus, the degree and type of joint adaptation may depend on the cost and gain of the cooperation outcomes as well as on other task-relevant information such as motor noise. These issues are particularly relevant to extending the joint adaptation model to a stochastic (i.e. probabilistic) model, and to understanding how the statistics about a task partner are established through the course of interaction.

6.6 Social Task Partner in Cooperative Action

In order to implement a specific reference trajectory for one of the pair, the studies from our lab that we described used a humanoid robot as a task partner. There is a continuing debate about the way people interact differently to movements of another person compared to non-biological agents. It is well known that people are tuned to detect biological motion and can extract a range of personal and psychological attributes from only a few points of markers representing movements of joints of a person such as the elbows and the shoulders [48–50]. Previously, it has been shown that believing one is coordinating an action with a human or a robotic partner can result in different behavioural responses by the perceiver regardless of whether the action is of human origin or a simulation of robot motion [51]. In addition, differences in neural responses have been demonstrated during observations of biological compared with non-biological movements, especially when the latter lacked the natural variability of human movements [52] or the former followed a non-Gaussian velocity profile [53]. In our study, although the robot executed minimum-jerk movements, which are known to closely approximate simple human arm movements, nonetheless the observer was fully aware from visual cues that he was interacting with a robot. Thus, caution should be exercised in generalising the conclusions regarding the results of this experiment to include natural social interaction between people. Nevertheless, it is interesting to explore how the adaptive behaviour described here is influenced by the social knowledge about the partner. For example, interpersonal interaction is known to be influenced by certain attributes such as emotional states [54, 55], relative physical characteristics [56] as well as the history of interaction [42]. We therefore believe that the approach we have presented can potentially provide a useful platform for investigating social determinants of interpersonal behaviour.

6.7 Conclusions

While researchers have generally tackled the challenge of understanding the human brain by focusing on a single actor interacting with a controlled environment, there is a growing appetite in neuroscience for investigating how the CNS operates at a social level. In particular, much attention is being paid to those studies of 'joint action' which are laying foundations for an understanding of how people represent conspecifics in the CNS [4]. However, there have been very few studies specifically focusing on the sensory-motor interactions between people. In this chapter we have described an approach to the study of joint action that focuses on the movements of the partners. In this manner we have described how the CNS learns and controls movement with respect to concurrent movement executed by a partner through feedforward and feedback control.

Acknowledgements   This work was partly supported by the ImmerSence project within the 6th Framework Programme of the European Union, FET—Presence Initiative, contract number IST-2006-027141, see also www.immersence.info and by BBSRC grant BB/F010087/1 to AMW.

References

1. Danchin, E., Giraldeau, L.A., Valone, T.J., Wagner, R.H.: Public information: From noisy neighbors to cultural evolution. Science **305**, 487–491 (2004)
2. Herrmann, E., Call, J., Hernandez-Lloreda, M.V., Hare, B., Tomasello, M.: Humans have evolved specialized skills of social cognition: The cultural intelligence hypothesis. Science **317**, 1360–1366 (2007)
3. Brosnan, S.F., Bshary, R.: Cooperation and deception: from evolution to mechanisms introduction. Philos. Trans. R. Soc. Lond. B, Biol. Sci. **365**, 2593–2598 (2010)
4. Sebanz, N., Bekkering, H., Knoblich, G.: Joint action: bodies and minds moving together. Trends Cogn. Sci. **10**, 70–76 (2006)
5. Sebanz, N., Knoblich, G., Prinz, W.: Representing others' actions: Just like one's own? Cognition **88**, B11–B21 (2003)
6. Tsai, C.C., Kuo, W.J., Jing, J.T., Hung, D.L., Tzeng, O.J.L.: A common coding framework in self-other interaction: evidence from joint action task. Exp. Brain Res. **175**, 353–362 (2006)
7. Knoblich, G., Sebanz, N.: The social nature of perception and action. Curr. Dir. Psychol. Sci. **15**, 99–104 (2006)
8. Rizzolatti, G., Sinigaglia, C.: The functional role of the parieto-frontal mirror circuit: interpretations and misinterpretations. Nat. Rev., Neurosci. **11**, 264–274 (2010)
9. Bosga, J., Meulenbroek, R.G.J.: Joint-action coordination of redundant force contributions in a virtual lifting task. Mot. Control **11**, 235–258 (2007)
10. Newman-Norlund, R.D., van Schie, H.T., van Zuijlen, A.M.J., Bekkering, H.: The mirror neuron system is more active during complementary compared with imitative action. Nat. Neurosci. **10**, 817–818 (2007)
11. Reed, K., Peshkin, M., Hartmann, M.J., Grabowecky, M., Patton, J., Vishton, P.M.: Haptically linked dyads—are two motor-control systems better than one? Psychol. Sci. **17**, 365–366 (2006)
12. Faisal, A.A., Selen, L.P.J., Wolpert, D.M.: Noise in the nervous system. Nat. Rev., Neurosci. **9**, 292–303 (2008)
13. Johansson, R.S., Westling, G.: Programmed and triggered actions to rapid load changes during precision grip. Exp. Brain Res. **71**, 72–86 (1988)
14. Darton, K., Lippold, O.C., Shahani, M., Shahani, U.: Long-latency spinal reflexes in humans. J. Neurophysiol. **53**, 1604–1618 (1985)
15. Wolpert, D.M., Ghahramani, Z.: Computational principles of movement neuroscience. Nat. Neurosci. **3**, 1212–1217 (2000)
16. Kawato, M.: Internal models for motor control and trajectory planning. Curr. Opin. Neurobiol. **9**(6), 718–727 (1999)
17. Thoroughman, K.A., Shadmehr, R.: Learning of action through adaptive combination of motor primitives. Nature **407**, 742–747 (2000)
18. Vorberg, D., Wing, A.: Modeling variability and dependence in timing. In: Keele, S., Heuer, H. (eds.) Handbook of Perception and Action, pp. 181–262. Academic Press, New York (1996)
19. Behrens, T.E.J., Hunt, L.T., Woolrich, M.W., Rushworth, M.F.S.: Associative learning of social value. Nature **456**, 245–249 (2008)
20. Kon, H., Miyake, Y.: An analysis and modeling of mutual synchronization process in cooperative tapping. J. Hum. Interface Soc. **7**, 61–70 (2005)
21. Repp, B.H., Keller, P.E.: Sensorimotor synchronization with adaptively timed sequences. Hum. Mov. Sci. **27**, 423–456 (2008)
22. Schuch, S., Tipper, S.P.: On observing another person's actions: Influences of observed inhibition and errors. Percept. Psychophys. **69**, 828–837 (2007)
23. Rabbitt, P.M.A.: Errors and error correction in choice-response tasks. J. Exp. Psychol. **71**, 264–672 (1966)
24. Notebaert, W., Houtman, F., Van Opstal, F., Gevers, W., Fias, W., Verguts, T.: Post-error slowing: An orienting account. Cognition **111**, 275–279 (2009)
25. Flash, T., Hogan, N.: The coordination of arm movements—an experimentally confirmed mathematical-model. J. Neurosci. **5**, 1688–1703 (1985)

26. Ito, S., Yuasa, H., Ito, M., Hosoe, S.: On an adaptation in distributed system based on a gradient dynamics. In: Proceedings of the 1999 IEEE International Conference on Systems, Man, and Cybernetics, Tokyo, Japan (1999)
27. Murian, A., Deschamps, T., Bourbousson, J., Temprado, J.J.: Influence of an exhausting muscle exercise on bimanual coordination stability and attentional demands. Neurosci. Lett. **432**, 64–68 (2008)
28. Griffiths, D., Tipper, S.P.: Priming of reach trajectory when observing actions: Hand-centred effects. Q. J. Exp. Psychol. **62**, 2450–2470 (2009)
29. Endo, S., Bracewell, R.M., Wing, A.M.: Learning cooperative object lifting using haptic feedbacks (submitted)
30. Quinn, M., Smith, L., Mayley, G., Husbands, P.: Evolving controllers for a homogeneous system of physical robots: structured cooperation with minimal sensors. Philos. Trans. R. Soc., Math. Phys. Eng. Sci. **361**, 2321–2343 (2003)
31. Vaughan, R.T., Støy, K., Sukhatme, G.S., Mataric, M.J.: Go ahead; make my day: Robot conflict resolution by aggressive competition. Anim. Animats **6**, 491–500 (2000)
32. Stone, P., Veloso, M.: Task decomposition, dynamic role assignment, and low-bandwidth communication for real-time strategic teamwork. Artif. Intell. **110**, 241–273 (1999)
33. Di Paolo, E.A., Rohde, M., Iizuka, H.: Sensitivity to social contingency or stability of interaction? Modelling the dynamics of perceptual crossing. New Ideas Psychol. **26**, 278–294 (2008)
34. Thompson, E., Varela, F.J.: Radical embodiment: neural dynamics and consciousness. Trends Cogn. Sci. **5**, 418–425 (2001)
35. Tognoli, E., Lagarde, J., deGuzman, G.C., Kelso, J.A.S.: The phi complex as a neuromarker of human social coordination. In: Proceedings of the National Academy of Sciences of the United States of America, vol. 104, pp. 8190–8195 (2007)
36. van Ulzen, N.R., Lamoth, C.J.C., Daffertshofer, A., Semin, G.R., Beek, P.J.: Characteristics of instructed and uninstructed interpersonal coordination while walking side-by-side. Neurosci. Lett. **432**, 88–93 (2008)
37. Zivotofsky, A.Z., Hausdorff, J.M.: The sensory feedback mechanisms enabling couples to walk synchronously: An initial investigation. Journal of Neuroengineering and Rehabilitation **4**(28) (2007)
38. Chartrand, T.L., Bargh, J.A.: The chameleon effect: The perception-behavior link and social interaction. J. Pers. Soc. Psychol. **76**, 893–910 (1999)
39. Wilson, M., Knoblich, G.: The case for motor involvement in perceiving conspecifics. Psychol. Bull. **131**, 460–473 (2005)
40. Richardson, M.J., Marsh, K.L., Isenhower, R.W., Goodman, J.R.L., Schmidt, R.C.: Rocking together: Dynamics of intentional and unintentional interpersonal coordination. Hum. Mov. Sci. **26**, 867–891 (2007)
41. Neda, Z., Ravasz, E., Brechet, Y., Vicsek, T., Barabasi, A.L.: The sound of many hands clapping—tumultuous applause can transform itself into waves of synchronized clapping. Nature **403**, 849–850 (2000)
42. Oullier, O., de Guzman, G.C., Jantzen, K.J., Lagarde, J., Kelso, J.A.S.: Social coordination dynamics: Measuring human bonding. Soc. Neurosci. **3**, 178–192 (2008)
43. Schmidt, R.C., Bienvenu, M., Fitzpatrick, P.A., Amazeen, P.G.: A comparison of intra- and interpersonal interlimb coordination: Coordination breakdowns and coupling strength. J. Exp. Psychol. Hum. Percept. Perform. **24**, 884–900 (1998)
44. Felmlee, D.H., Greenberg, D.F.: A dynamic systems model of dyadic interaction. J. Math. Sociol. **23**, 155–180 (1999)
45. Vorberg, D., Schulze, H.H.: Linear phase-correction in synchronization: Predictions; parameter estimation; and simulations. J. Math. Psychol. **46**, 56–87 (2002)
46. Endo, S., Wing, A.M., Diedrichsen, J.: Joint adaptation in cooperative action (submitted)
47. Braun, D.A., Ortega, P.A., Wolpert, D.M.: Nash equilibria in multi-agent motor interactions. PLoS Comput. Biol. **5**(8), 1–8 (2009)
48. Brownlow, S., Dixon, A.R., Egbert, C.A., Radcliffe, R.D.: Perception of movement and dancer characteristics from point-light displays of dance. Psychol. Rec. **47**, 411–421 (1997)

49. Blake, R., Shiffrar, M.: Perception of human motion. Annu. Rev. Psychol. **58**, 47–73 (2007)
50. Shim, J., Carlton, J.G.: Perception of kinematic characteristics in the motion of lifted weight. J. Mot. Behav. **29**(2), 131–146 (1997)
51. Stanley, J., Gowen, E., Miall, R.C.: Effects of agency on movement interference during observation of a moving dot stimulus. J. Exp. Psychol. Hum. Percept. Perform. **33**, 915–926 (2007)
52. Gazzola, V., Rizzolatti, G., Wicker, B., Keysers, C.: The anthropomorphic brain: The mirror neuron system responds to human and robotic actions. NeuroImage **35**, 1674–1684 (2007)
53. Kilner, J.M., Paulignan, Y., Blakemore, S.J.: An interference effect of observed biological movement on action. Curr. Biol. **13**, 522–525 (2003)
54. Grammer, K., Kruck, K., Juette, A., Fink, B.: Non-verbal behavior as courtship signals: The role of control and choice in selecting partners. Evol. Hum. Behav. **21**, 371–390 (2000)
55. Janovic, T., Ivkovic, V., Nazor, D., Grammer, K., Jovanovic, V.: Empathy, communication, deception. Coll. Antropol. **27**, 809–822 (2003)
56. Isenhower, R.M., Marsh, K.L., Carello, C., Baron, R.M., Richardson, R.M.: The specificity of intrapersonal and interpersonal affordance boundaries: intrinsic versus extrinsic metrics. In: Heft, H., Marsh, K.L. (eds.) Studies in Perception and Action, VIII, pp. 54–58. Erlbaum, Hillsdale (2005)

Chapter 7
Cooperative Physical Human-Human and Human-Robot Interaction

Kyle B. Reed

Abstract This chapter examines the physical interaction between two humans and between a human and a robot simulating a human in the absence of all other modes of interaction, such as visual and verbal. Generally, when asked, people prefer to work alone on tasks requiring accuracy. However, as demonstrated by the research in this chapter, when individuals are placed in teams requiring physical cooperation, their performance is frequently better than their individual performance despite perceptions that the other person was an impediment. Although dyads are able to perform certain actions significantly faster than individuals, dyads also exert large opposition forces. These opposition forces do not contribute to completing the task, but are the sole means of haptic communication between the dyads. Solely using this haptic communication channel, dyads were able to temporally divide the task based on task phase. This chapter provides further details on how two people haptically cooperate on physical tasks.

7.1 Introduction

There are many ways to classify human-human interaction. Two people can interact by speaking, changing facial expressions or body posture, shaking hands or hugging, and written word. Some types of human-human interactions have been studied extensively, such as interactions at a distance. Simply by sight alone, two people will naturally and subconsciously synchronize their actions, such as swinging a leg [1], and are able to consciously synchronize a swinging pendulum [2, 3]. One explanation for this ability is that mirror neurons in the brain can develop a representation of actions performed by another individual [4, 5]. In another study, Sebanz et al. [6] show that two participants working in close proximity to each other on different tasks can influence each other.

Although there are significant interactions that occur at a distance, the research discussed in this chapter focuses on how groups of individual agents *physically* work together. The physical interaction between two people directly connected has

K.B. Reed (✉)
University of South Florida, 4202 E. Fowler Ave, ENB118, Tampa, FL 33612, USA
e-mail: kylereed@usf.edu

A. Peer, C.D. Giachritsis (eds.), *Immersive Multimodal Interactive Presence*,
Springer Series on Touch and Haptic Systems,
DOI 10.1007/978-1-4471-2754-3_7, © Springer-Verlag London Limited 2012

only recently been studied with significant rigor even though groups of people have been working together throughout history, for example during tug-a-war and when a giver knows when a receiver has control of a drinking glass and can let go [7]. In a review of joint action, Sebanz et al. [8] suggest that understanding how groups of people interact will likely further our understanding of how the brain works in isolation and state that "the ability to coordinate our actions with those of others is crucial for our success as individuals and as a species." However, complications can arise in physical communication since a person perceives self generated forces and received forces differently [9]; each person may believe the other person is more commanding than they really are.

In one of the first studies on human-human interaction, Wegner and Zeaman [10] discussed some control tasks that occur between multiple individuals in everyday life, such as the see-saw, the two-handled saw, and balancing on a tandem bicycle. Shaw et al. [11] also discussed couples teaching physical activities, such as swinging a golf club and dancing. Other common activities requiring cooperative control are moving and placing large objects, exchanging objects like a glass of water without spilling, and symmetrically positioning a bed linen on a mattress. In these examples, the two people develop a cooperative partnership in which they must divide control and compromise according to the task at hand. Knoblich and Jordon [12] suggest that group coordination may be beneficial since each person in the group has fewer actions to deal with.

Devices that mediate the interaction between two people, such as teleoperators and compliant structures, often inhibit physical communication. When two people are working together, it is important that each member feels the force from the other person or object. Many haptic devices cannot reproduce the forces perfectly, which makes interaction through the devices particularly difficult. Force reproduction can become a significant problem when working over great distances, such as in teleoperation. Teleoperation research tends to look at issues related to accurately recreating forces, such as time lag, and less on how the two remote agents interact. The work discussed in this chapter focuses on the cooperation between two or more agents with essentially no time delay and a high fidelity interaction. The goal of many of these works is to understand the fundamental interactions for later inclusion into other robotically mediated interactions.

Before looking at human interaction further, it is interesting to look at a species that is also highly adept at physical cooperation. Some of the effects seen in groups of humans interacting can be seen in groups of ants, specifically in the Asiatic Ant (*Pheidologeton diversus*). In a paper outlining how ants transport food in groups, Moffet [13] wrote: "Group transport (the carrying or dragging of a burden by two or more individuals) is better developed in ants than in any other animal group." The most notable finding from his study is that each ant working in a group carries more weight per ant than an ant carrying an object alone. One ant alone can carry five times its body weight. Yet, in one example, 100 ants worked together to carry a worm that weighed 5000 times the body weight of an ant; each ant carried 50 times its own body weight. Moffet found that the ants can carry exponentially more weight with increasing ants until about 11 ants are working together. Moffet speculated that the space around the perimeter of an object limits the effectiveness of ants that can actually work together, thus group effectiveness grew slower for groups larger than 11 ants.

Moffet also found that transport velocity was fairly constant for groups of 2–10 ants, which were twice as fast as individual ants. Groups of 11 or more slowed down to less than half the speed of smaller groups. His efficiency metric of velocity × object weight showed that the ants were increasingly efficient as more ants worked together up to groups of 11 ants, at which point effectiveness increased at a slower rate.

Not all ant species use cooperative group transport. In other ant species, if a second worker ant grabs an object, the first ant's progress is halted until the second worker releases the object. These species tend to break the food down into smaller pieces and carry them individually, which could be thought of as the most rudimentary form of group cooperation. It is only some species of ants that have developed this ability to cooperatively carry large work loads.

Although ants cooperating to carry large loads is not directly related to studying how humans interact or how a robot can cooperate with a human, there have been several studies that used ants and other insects as inspiration for developing swarms of smaller robots to work together [14, 15]. These bio-inspired robotic studies have had some success and are likely to benefit from a further understanding of group dynamics.

The body of literature presented in this chapter aims to bring together the recent studies on the cooperative motion of groups of humans as well as groups of human-robot teams. The chapter starts with the performance of groups compared to individuals in Sect. 7.2, then the interaction forces between the two individuals are discussed in Sect. 7.3, and finally, several methods of implementing human-robot interaction based on how two humans interact are discussed in Sect. 7.4.

7.2 Group Performance

It is often accepted that people prefer to work alone on tasks that require accuracy, finding a partner to be an impediment. However, in one of the first studies on cooperative motion, Wegner and Zeaman [10] found that dyads could follow a path significantly better than individuals, and quads significantly better than both dyads and individuals. To compare groups to individuals working alone, they used a "pursuit rotor" task in which a participant tried to follow a path marked on the top of a rotating turntable. However, they were unable to determine a satisfying explanation, possibly since they did not measure the forces exerted by the participants. This section will first focus on the nature of group performance, how perception affects group performance, and a discussion about how Fitts' Law applies to multiple physically interacting humans; the interaction forces will then be described in Sect. 7.3.

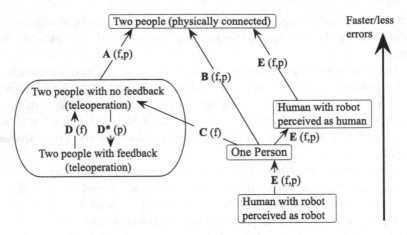

Fig. 7.1 Summary of several experiments that compare the speed and error rate of several algorithms for mediating human-robot-human interactions. Each of these used a second order inertial system, except for D*, which used a zero-order system. An 'f' indicates that the other member's force was displayed while a 'p' indicates position was displayed (**A**—Summers et al. [16] and Field et al. [17]; **B**—Reed et al. [18, 19]; **C**—Glynn et al. [20]; **D**—Glynn et al. [21]; **E**—Reed et al. [22, 23])

7.2.1 Human-Human Group Performance

A number of recent studies have examined the performance of different group compositions and the effect that the interface between the members has on the performance. Figure 7.1 summarizes many of these results and the remainder of this section will expand on each of the studies referenced in the caption.

Glynn and Henning [20] examined several methods of combining the forces from two members of a dyad. They showed that using the average of the commands without haptic interaction resulted in faster and more accurate task execution than one person alone. In their maze following study, the average of the force applied by each partner to a joystick controlled the acceleration of an inertial mass. The participants were not physically interacting as there was only visual feedback and no force feedback. The force applied to the virtual mass was the average of the desired path of each operator who cooperated without haptic interaction. Glynn and Henning found that teams completed the maze 14.5 seconds faster than individuals, an increase of 8%. Collisions were reduced from 484 for individuals to 334 for teams. It makes sense that the collisions are less since any faulty motion would be diluted by the other member's action. Statistically, the chance of both participants producing the same collision-producing motion would be less than that of one person acting alone. The improved performance may also be explained by a diluting effect. Since the action of each participant is diluted, the effect may not be as noticeable, and thus, both participants may attempt to push harder than they would alone. This could result in each participant applying more force and, thus, completing the task faster. Since there was no haptic interaction between the participants, this effect would not likely have been noticed.

A related study by Summers et al. [16] examined similar methods to mediate the input commands from two people using the two flight sticks in an airplane cockpit. Just like Glynn and Henning's study, they found a significant performance benefit when using coupled interaction compared to the uncoupled Fly-By-Wire method. The uncoupled Fly-By-Wire is essentially averaging the inputs from each flight stick and the coupled interaction is similar to the mechanical method where both sticks have the same motions. Even though performance is better for two people with an average of their commands, it does not necessarily mean that pilots, or other shared control tasks, should use this control mechanism as this could have detrimental effects; imagine one pilot pulling up and the other pushing down to avoid a collision—the average would be straight ahead. The forces, which will be discussed in the next section, are an important part of mediating a physical interaction.

In an extension to the maze studies, Glynn et al. [21] also used force feedback, so the participants were haptically interacting. The interaction was simulated using a spring between the two manipulandums. They compared the interactions with and without force feedback using position and force control. The added feedback when using position control improved performance. During force control, the performance time was unchanged. With two people interacting physically, they are able to communicate both on position and force, thus there should be no detrimental interaction like there is in the force control experiment.

In a series of experiments using a 1 DOF crank with two handles, Reed et al. [18] examined the completion times of dyads performing a target acquisition task with a rigid connection; the forces and motions were directly conveyed to each individual. To allow comparison of dyads and individuals, the experiment consisted of one or two participants completing the task of moving a crank into a series of targets. During the experiment, many of the participants reported the typical opinion of difficulty when working with a partner; few reported cooperation. This perception of poor cooperation likely stems from the increased force exerted by each member of the dyad. In fact, each dyad member applied forces 2.1 times larger than when working individually. Most of this increased force was applied in opposition to the other member.

Despite the increased forces and lack of perceived cooperation, dyads completed the task faster than individuals [18]. The completion time for dyads decreased by an average of 54.5 ms compared to the average completion time of the two constituent individuals working alone. The average completion time for individuals was 680 ms. Increased force associated with the dyad condition might result from a faster participant pulling along a slower one. However, dyads averaged 24.8 ms faster than the faster of the constituent individuals working alone. Only two of 30 participants were faster than their respective dyad. The dyads established the faster performance quickly, generally within 20 seconds after the dyad started working together. The rotational inertia of the crank was doubled in the dyad case, so the faster performance cannot be caused by sharing the load between the dyad members.

7.2.2 Perception Affects Performance

The performance of individuals in a group setting is often correlated with the perceived environment around them, such as who is watching or what they are interacting with. To examine the effects of the perception of one's partner, Reed et al. [18, 19] extended their original crank study by recreating a partner using the forces found during their previous human-human experiments. In order to prevent the variations in forces and completion times among participants from affecting the results, the robotic partner was based on a recorded version of the individual's forces during individual trials. Since the performance was based on each individual, any differences can be attributed to the interaction of the simulated partner and not because the robotic partner was faster or slower than the participant. The participants were expected to work with the robotic partner in a similar way as they did with a human partner. The interaction forces are further described in Sect. 7.3.1.

In this experiment, each participant performed the crank task both individually and with the robotic partner. Half the participants did not have a human on the other side of the table, so they clearly knew they were working with some non-human agent. The other half of the participants had a human across the table and assumed they were working with a human. A curtain prevented the participants from knowing what the other person was actually doing. The human "confederate" was actually working with the experimenter and did not actually participate to move the crank; the robot did. Ten of the eleven participants with a confederate present stated that they thought a person was working with them and were surprised to discover at the end of the experiment that they worked with a robot and had not worked with the other person. Thus, the participants working in the presence of a confederate consciously believed they were working with a human partner. This indicates that such a robotic partner was able to cognitively pass a haptic version of the Turing test [24–26].

When a confederate was present, the human working with the simulated partner was on average 5.8 ms (1%) faster than the same participant working alone. The human-human teams were 48.8 ms (7%) faster than the human-robot team with a confederate present. When the confederate was not present, the participant working with the robot was 24.8 ms (3.5%) slower than working alone. When the participants were aware that they were working with a non-human agent, the participants performed worse than when they were working alone. When the participants thought they were working with a human, their performance did not change relative to their individual performance. This implies that the perceived origin of forces in physical collaboration affects how a person will interact with a partner.

One hypothesis for why dyads are faster is social facilitation [23]. Social facilitation research has a long history [27, 28] with many studies showing that simply having a person in a room observing a participant will lead to better performance on a given task. This is typically explained by the mere presence of others elevating drive levels. Mere presence tends to improve performance on simple, or well mastered, tasks and inhibits the performance on complex, or poorly mastered, tasks [29]. Social facilitation may have accounted for some of the performance increase. However, the participants knew the experimenter was always watching, so, in both versions of the test, there was someone visually present, which is the sole requisite for improved performance as stated by social facilitation. Wegner and Zeaman [10], when studying groups and singles on a pursuit rotor task, suggested social facilitation as an explanation of the increased performance of groups, but stated that it could not account for all of their observed effects. Also, during the cooperative crank studies, the participants could only feel the other member of the dyad and they could not see each other due to a curtain hanging between them. Social facilitation has only been demonstrated through visual interaction, not physical interaction. The only aspect of the task that is changing is whether they are holding the handle or not. It is possible this physical change could elicit the same effect, but in all other versions of social facilitation described in the literature, the two participants are physically disconnected from each other. They are predominantly communicating through vision, which is the basis of most of the social facilitation literature.

This performance increase is likely due to the motor control systems of both people working cooperatively and not for the reasons explained in the social facilitation literature. The improved performance suggests that social facilitation could also be caused by a similar "haptic presence" effect, but this needs further evaluation in future studies. Even if haptic presence had a similar effect to social facilitation, it would not be able to fully explain all of the improved performance for dyads on the cooperative crank tasks. Dyads improved by 54.5 ms, whereas the difference of performance increase between the human-robot and human-robot-confederate groups was only 24.8 ms, half the difference time.

7.2.3 Groups Obey Fitts' Law

Fitts' law [30] is an empirical relation observing that the time it takes a person to reach a target is related to the distance to the target, D, and the size of the target, S. The distance and target size vary linearly with the index of difficulty ID, which is defined as the logarithm of $\frac{D}{S}$, or $t = a + b * \log_2(\frac{D}{S})$, where a and b are constants specific to the task. In other words, for a given target size, it takes longer to move a large distance than a small distance and, for a given distance, it takes longer to move into a small target than into a large target.

Fitts' law has proven to be remarkably robust since it was first described in 1954. It has been used to describe and analyze tasks of varying complexity and in multiple degrees of freedom, for instance: path tracking [33]; 3D computer games [34]; scrolling time on a computer [35]; GUI design [36]; cursor movement along a line to a target line segment; cursor movement in a plane to a target disk, stringing beads, placing a can on a shelf, putting a peg into a hole, and many others [31]. The coefficients of Fitts' law vary depending on the task (and from person to person), but the linearity in $\log(\frac{D}{S})$ is observed throughout a wide range of the index of difficulty measure, and over a wide variety of tasks, which makes it well suited to analyzing how two people cooperate on a physical task. Fitts' law has recently been tested for multiple people interacting.

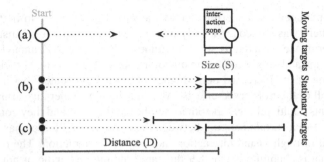

Fig. 7.2 The variations of Fitts' Law for two people. (**a**) The target is the other person's hand, so the target is moving toward as each member of the team attempt to reach the target. This type of experiment was performed in [7, 31]. (**b**) The target is stationary and the two members work together to reach the same target [32]. (**c**) The two individuals have different, but overlapping, targets, so they must physically work together to reach the target [32]

Sallnäs and Zhai [7] used Fitts' Law to study how two individuals handoff an object in a haptic simulation. Two people sat in-front of a computer with a manipulandum to control the position of their virtual hand. The size of the interaction zone varied according to Fitts' law. They measured the time it takes for one person to hand off a virtual object to another person within certain spatial targets (see Fig. 7.2a). Sallnäs and Zhai found that Fitts' law is valid for a two person handoff task. They performed experiments with and without force feedback in the manipulanda. They found that performance time did not change significantly with different amounts of force feedback, but the error rate (number of dropped objects) was significantly lower with haptic feedback. The two people in this study only felt the sensations the haptic device can simulate since they were not physically connected.

The experiment performed by Sallnäs and Zhai's consisted of participants exchanging an object within certain spatial limits. In everyday life, these spatial limits do not exist. In this case, the target would be moving, not stationary, as shown in Fig. 7.2a. Mottet et al. [31] performed a Fitts' law study with moving targets. The setup consisted of two manipulanda and two displays of LEDs each showing the position of their target and the position of their own manipulandum. Each person's target was the other person's manipulandum. As each person moved towards the target, the target also moved closer to them. Mottet et al. showed that this type of dual motion task also obeys Fitts' law, meaning that for small targets, it takes longer to reach each other than it does for larger targets.

When two individuals are physically cooperating on the same task, their performance time also obeys Fitts' law [32], but does not when they must compromise. In this experiment, two individuals were cooperatively moving a crank into targets. Two types of targets displayed to the users were tested, as shown in Fig. 7.2b and c. In one case, the targets shown to each member were the same; the second case showed different, but overlapping, targets to each member. In the same target case, Fitts' law was obeyed. In the different overlapping target case, the performance did not obey Fitts' law regardless of whether the individual target size or the overlapping target size was used. The deviation from Fitts' law is likely caused by the necessary compromise since the only solution required each member to be on the edge of the target, not in the center as they were when the targets were the same size.

7.3 Force Interactions

The forces that each member of a group feels during an interaction allows them to determine many aspects about the object and the person. For example, one can quickly get a distinct impression from the firmness of a handshake and people rarely drop objects when exchanging them because they know when the other person has control of the object. Prior to studying the interaction forces, several researchers were unable to adequately explain the effects felt when cooperating on a task. Wegner and Zeaman [10] reported that some of the participants completing a "pursuit rotor" task mentioned that the mechanism felt more mechanically stable in group conditions. The participants in this study manually controlled a stylus via a handle in two dimensions. The stylus had multiple handles so that individuals or teams of two and four could be tested in the same way. The investigators tried to increase the stability through mechanical means, but their attempts only decreased the path tracking ability. Many of their attempts to understand the interactions were unsuccessful since they did not measure the forces applied by each of the individuals. This section will look specifically at several recent studies of the forces involved between members of a group when completing various tasks.

Shergill et al. [37] examined the forces exchanged between two people without motion. In this experiment, two participants each put their finger in a lever attached to a force transducer. Each participant was told to push with the same force they felt, but the participants were unaware of the instructions given to the other participant. Alternately, as instructed, each participant applied the force. Shergill et al. found that in every case, the forces escalated from trial to trial. They explained that the participants are reporting the true perception of the force and the increasing force is due to neural processing.

In a second set of experiments, Shergill et al. asked the participants to recreate a force applied on one finger with a finger from the other hand. The participants consistently generated a force larger than the original. Shergill et al. suggest that externally generated forces are perceived as stronger than internally generated forces. This result implies that each person in a group will always feel that he/she is contributing less to the overall task than the other member, even though that is not and cannot be the case. When working cooperatively with a partner on a task, each participant may want to contribute equally, which could lead to an escalation in performance.

In some applications, forces can relay vital information. If the perception of forces is reduced, as suggested by Shergill el al.'s study [37], or the transfer of forces is hindered, communication can be significantly diminished. Fly-By-Wire (FBW), a design for airplane control, eliminates the direct mechanical connection (and thus some of the forces) between the pilot and the plane's control surfaces and also between the two pilots. Depending on the configuration and design of the FBW system, the flight sticks allow little or no haptic interaction between pilots. Summers et al. [16] conducted a series of experiments on pilots using a Flight Simulator at NASA Ames Research Center. They examined four different cases, ranked by the pilots in order from most preferred to least preferred: coupled, uncoupled with a disconnect switch, uncoupled with priority logic (essentially the largest input wins), and uncoupled (average of inputs). The pilots significantly preferred the coupled (haptic) FBW more than the uncoupled (non-haptic) FBW.

In a study among 157 commercial aircraft pilots, Field and Harris [17] found that communication was lost when using FBW. This loss of communication is likely to decrease the pilots' awareness of current situations. Many pilots in the study stated that it is important and useful to be able to feel the forces and motions of the other pilot and that they wanted to feel what the autopilot was doing so they could determine if the plane was flying correctly. In a direct mechanical connection, the pilots can feel what the other person is doing as well as a response from the plane itself. Shergill et al. [37] explained how an individual can separate their forces from an external force, but the pilots are unable to separate the forces in a flight stick into those from the other pilot and those from the force feedback of the plane.

Glynn et al. [21] performed experiments where participants had to jointly track an object. The haptic display had force feedback that modeled the interaction as a spring, so the two participants could feel the motions and forces of the other person. The force feedback displayed to each person was programmed as either "social force-feedback" where each dyad member could sense the position of the other dyad member via force feedback and "system force-feedback" where both dyad members could sense the simulated mass. They compared these two conditions and the same two conditions with a 0.25 second lag and a no force feedback condition. The lag increased both time taken and collisions. Without force feedback, the participants caused less damage and were able to stay closer to the center of the path. Glynn et al.'s only explanation for the larger path deviation was that the feedback interacts with the dynamics of the second order system in complex ways. Just like the pilots in Field and Harris's study, it is possible that each member has difficulty separating the force feedback of the device from the other member's forces. The force feedback is possibly more beneficial during position control since the position does not overlap with the force. With two people interacting physically, they are able to communicate both on position and force, thus there should be no detrimental interaction like there is in the force control experiment.

Another cooperative feedback mode was discussed and compared to two cooperative feedback modes similar to those in Glynn et al.'s, shown in Fig. 7.3. In this experiment, the interaction between the two humans and a slave robot were mediated by three control laws designed for mediating different aspects of the redundant control afforded by multiple human users [38]. The third mode, called Dual Force Feedback, allows both users to feel the same force that is proportional to the difference between the average position of the two master robots and the position of the slave robot. This has the advantage that both participants feel the exact same force. However, it has the disadvantage that it is nearly impossible to distinguish whether you are fighting with your partner or if the slave robot is restricted. This mode was

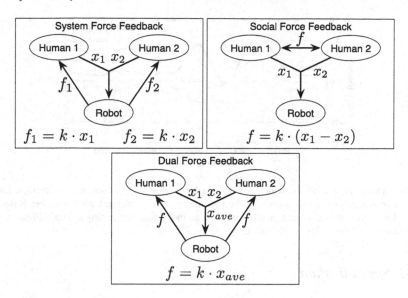

Fig. 7.3 Figure showing the three interaction modes used in [21, 38]. Each mode mediates the redundancy between the two users to enable different types of interaction

found to cause more fighting between the two participants, but also allows identical forces to be felt by each user, which would be beneficial during teaching/learning tasks.

Although couples dancing has been mentioned as an example of a physical interaction, it had not been studied until Gentry [39, 40] examined how this interaction works between dancers. She describes dancing as a finite state machine where dancers move between a limited number of poses and interact through force and motion. Dancers coordinate their actions through various elements, some internal to the dancers and some external. The rhythm, or beat, of the music is an external event that synchronizes the motion of each dancer's movements and is the strongest element coordinating their actions. The poses are a predetermined position of the bodies of the dancers. Some dances, such as the Waltz, have only one pose, whereas Lindy Hop has multiple. The leader moves from position to position while indicating what to do next. The transitions are coordinated based on previous knowledge of a small set of moves. In another study using haptic recordings of couples dancing, a male dancing partner was synthesized [41]. An adaptation law allowed the step size of the robot to change to accommodate the female partner.

The physical connection in dancing is maintained though the follower's right hand holding onto the leader's left hand. This physical connection allows the leader to send messages to the follower as well as for both partners to exchange energy. A good follower will keep her hand in the same position relative to her body, which allows the leader to communicate. Most of the communication is based on haptic cues even though the dancers can see each other. Gentry performed experiments on experienced couples dancing blindfolded and found that they were capable of performing quite well with only haptic communication.

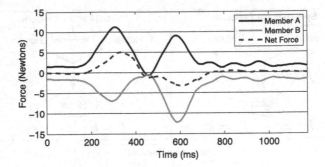

Fig. 7.4 The dyad specializes, which is where each member only pushes in one direction. Each member contributes to certain aspects of the task: member A accelerates while member B decelerates. The forces sum to a force profile similar to an individual performing the task. Figure used with permission from [23] (© 2007 IEEE)

7.3.1 Specialization

When a single person becomes part of a dyad, many new solutions to completing the task develop due to redundant limb motion [42, 43]. There is no longer a one-to-one correspondence between dynamics and kinematics. For example, in a dyadic task, one member of the dyad can choose not to contribute at all, to only help with the acceleration phase, or only to use elbow extensor muscles. Knoblich and Jordon [12] suggest that when groups work together, they might perform better because each person has fewer actions to deal with. In essence, they hypothesize that the members can get out of each other's way and only deal with a few actions.

In a series of experiments to further understand how two humans resolve the redundancy problem occurring during cooperative motion, Reed et al. [44] showed that two people naturally specialize their forces. They analyzed the forces from each individual on a one DOF target acquisition task. They transformed the forces exerted by each individual into a "net force" and a "difference force". The net force is the sum of the members' forces, which is the task relevant force that accelerates the crank. The difference force has no physical effect on crank acceleration and is a measure of the disagreement between the two members. A similar measure that excludes forces with different magnitudes in the same direction has also been used to quantify the interaction forces [45].

Figure 7.4 shows the net, difference, and each member's force for a single trial by a dyad exhibiting the typical specialization pattern. The dyad completed this task at approximately 600 ms. Even though the constituent members of a dyad produced very different force profiles, the net force for dyads produced a trajectory similar to the minimum jerk trajectory that individuals performing alone typically produce on reaching tasks [46]. The members of a dyad divide the task to achieve the same motion as an individual. As shown in Fig. 7.4, member A pushes toward the target during the beginning of the trial to accelerate the crank while member B either passively or actively resists the acceleration. Member B then pulls away from the target during the later part of the trial to decelerate the crank while member A continues applying a force toward the target. During the entire trial, member A primarily contributes to accelerating the crank and member B primarily contributes to decelerating the crank. This acceleration/deceleration specialization pattern is clearly revealed by inspecting the difference force, which remains always the same sign. Only the characteristic shape of the difference force matters—it could be mirrored around the x-axis if the participants were standing on opposite sides of the crank.

A metric for measuring specialization within a dyad is based on the contribution from each participant for each phase of the task. A highly specialized dyad would consist of one member contributing to most of the force during the acceleration phase and the other member primarily contributing to the deceleration phase. To find the contribution from each dyad member, the forces applied to each handle over the acceleration and deceleration phases were integrated and divided by the integrated net force for that phase. The result provides four fractional contributions of each member of the dyad during the acceleration and deceleration phases. The contributions during each phase from both members of the dyad necessarily sum to one. A negative contribution indicates that the member was applying a force opposite to typical motion of that phase, for example accelerating during a deceleration phase, or decelerating during an acceleration phase, even if the force was only due to passive inertia. A contribution greater than one indicates that this member had to apply a large force to compensate for the negative contribution of the other dyad member. As a comparison, specialization could also occur in another way where one member always pushes right and the other always pushes left, which would be expected for a left-handed member paired with a right-handed partner. In Reed et al.'s study [22], eleven of the fifteen dyads show significantly more acceleration/deceleration specialization than left/right specialization.

The dyads in the crank task learned to specialize their applied forces temporally to generate a net force similar to how a single person would complete the task, but faster. The participants divided the task solely through a haptic communication channel as no other communication was allowed. Similarly, Feygin et al. [47] found participants could learn the temporal aspects of tasks better using haptic guidance than they could using visual guidance. One hypothesized way specialization could be implemented in the human control system is to precue an action [48] so that when some other event happens they perform a certain prepared action [19]. With two people cooperating on a task, the accelerator could focus on the start of the task and the deceleration specialist could wait for some cue, such as reaching a particular location and/or velocity and would then begin to decelerate the crank.

A person individually performing a target acquisition task would be expected to use the triphasic burst pattern where an agonist muscle burst initiates the movement and an antagonist muscle burst is initiated to brake the movement and a second agonist burst is initiated to maintain the limb at the final position [49–51]. These bursts represent careful planning based on prior knowledge, rather than feedback received during the task [52, 53]. These patterns have been shown to represent optimal movements that accomplish the task within limiting physiological constraints such as the muscles activation rates [53, 54] and the limited torque [55] and force [56] generating capacity in different areas of the workspace. The specialization of roles has also been shown to be beneficial from an energy flow analysis [57]. Consequently, the rate at which muscles can turn on and off becomes a less limiting factor if one person can be ramping up while the other is ramping down, which is presumably what specialization enables the two members to accomplish.

In another study examining specialization on a 1 DoF crank system with two people, the radial force, which is the force directed toward the center of the crank that does not contribute to any acceleration or deceleration, was also studied [58–60]. They found that the radial forces were larger than the tangential forces in many instances. One of their conclusions was that the radial force stabilizes the interaction around a certain set point, much like the restoring effect that gravity has on a pendulum, but where the individual's radial force serves as the restoring force.

An individual performing a bimanual task exhibits a similar specialization strategy. Reinkensmeyer et al. [61] show that an individual holding a pencil between two fingers on different hands will use one hand to accelerate and the other hand to decelerate the object, which might be taken as a bimanual model for this observed two person acceleration/deceleration specialization. For a single individual the inward force from both hands allows the pencil to accelerate and decelerate while being tightly held and not dropped. The tight neural coupling between both arms allows an individual to effectively coordinate the actions of each arm [62–64]. However, in dyadic tasks there is no neural coupling between the individuals, so the developed strategy must have occurred through the haptic communication channel instead since the participants could only communicate physically.

7.3.2 Perturbation Rejection

Individuals are able to adapt to perturbations from the environment, either from unexpected deviations in a trajectory or from an external source. One method of overcoming external forces is to co-contract both the agonist and antagonist muscles on the same joint, which increases the stiffness in individuals [65, 66]. Co-contraction is a common strategy when individually interacting with unstable force fields [67, 68]. If a person is interacting with an ungrounded object individually, any force applied to the object will cause that object to accelerate. When working with a partner, each member can apply a force even though the object is not accelerating, as long as the sum of the forces are equal and opposite.

At the end of each trial on the crank interaction tasks discussed in Sect. 7.3.1, the two members bring the crank to rest as they wait for the next target to appear. During this time, the dyad members exert an average of 4 Newtons of force in opposition to one another [23]. This force may help dyads to resist perturbations by increasing the stiffness of the dyad in the same way that muscle co-contraction increases arm stiffness in an individual. In dyads, the force applied in opposition to each member could serve this same purpose. Since this type of interaction is a type of co-contraction within a dyad, it has been called "dyadic-contraction" [23]. Dyadic-contraction is a strategy similar to those used in parallel robotics and in human bimanual control [69, 70].

The dyadic-contraction force was examined by looking at individuals and by looking at dyads [22]. In the individual case, an external force of 0, 5, and 10 Newtons was applied to the participant as they performed a target acquisition task. At the end of each trial, a motor applied a 100 ms perturbation force and the change in position due to the perturbation was used as a measure of the perturbation effect. The three force levels showed a significantly different response with the largest displacement at the lowest force and the smallest displacement at the largest force [22]. This shows that an externally applied force can act in a similar fashion as co-contraction to help stiffen a person's arm.

The location of the arm when the perturbation was applied also had a significant effect on the response. Part of this effect was due to the arm reaching across the table so the inertia ellipse changed [71, 72]. The other effect was due to the direction of the perturbing force. On the participant's dominant side (i.e., same side as the hand holding the crank), the displacements were larger when the perturbation pushed away from the center of the crank.

Unlike the performance metrics, dyads actually performed this perturbation task worse than they did as individuals [22]. A similar pattern of perturbation rejection was found in same-handed dyads as in individuals, but the dyads had larger displacements. In many of the previous experiments, handedness did not make a difference, but because the inertia ellipses were different for left and right handed individual, it did make a difference here. When the dyad members consisted of one left-handed and one-right handed member, the overall perturbation rejection characteristics were better than the average of all locations of the same-handed dyads, but not quite as good as the best of the same-handed dyads. When the dyads had slightly different roles, the ability to reject a perturbation improved and was similar to the average of an individual. Figure 7.5 shows the results of the dyadic crank perturbation study.

Another hypothesis is that dyadic-contraction could also serve as a simple message between partners that they are working with a partner [22]. Without applying any force, there is no way to know what is happening or who is on the other side of the curtain. By applying a small force, each person feels a resistance that helps them to determine what is opposite them. In order to explore a surface, a person may try to maintain an optimum force [73, 74]. In order to learn about their partner, each member may be trying to keep the optimum force.

The question of what measures are correlated to other measures often arises when looking at how two humans interact. In [75], many of these correlations are examined; specifically the reaction times of the individuals, which dyad member started first, how much specialization was present, the performance of the dyad, and the amount of the dyadic-contraction force. All but two of these correlations had an R^2 value of 0.1 or less, so they do not indicate much of a relationship. It would be reasonable if the member with the faster reaction time on the individual trials would be the member that would be the acceleration specialist. However, the individual with a faster reaction time only had a correlation of $R^2 = 0.25$ with being the accelerator. The highest correlation was the dyadic-contraction force compared to the dyad member that started the motion, which had an $R^2 = 0.73$. This is likely because the participant that is already pushing toward the target due to dyadic-contraction tends

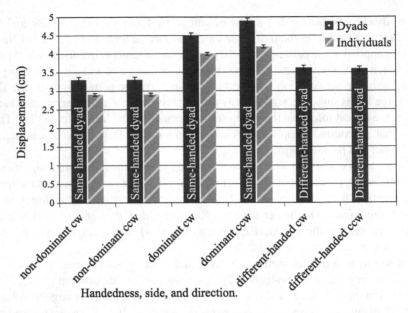

Handedness, side, and direction.

Fig. 7.5 The individuals and same-handed dyads exhibit the same pattern, but the dyads had a larger response to the perturbation. The same-handed dyads did not have improved performance by cooperating like they did for specialization. However, the different-handed dyads exhibited a consistently good cooperative effort as they appear to use the best of each member's ability for each perturbation type. Error bars represent 95% confidence intervals. Figure used with permission from [22] (© 2007 IEEE)

to push in the direction of the new target. This likely relates to the different ramp up/down characteristics of muscles [76].

7.4 Human-Robot Cooperation

One of the goals of studying human-human interaction is to further our understanding of how humans fundamentally work with another agent. Several groups have attempted this feat to varying levels of success.

Since two participants can perform a strategy similar to a single person performing a bimanual task, would a simulated partner that displays an acceleration/deceleration specialized trajectory be enough to elicit the same response? Reed et al. [22] examined this question using a simulated partner in a test similar to the Turing Test [24]. Alan Turing originally proposed a test for evaluating a computer's ability to produce human like conversation via text, but it has since been used more broadly for other human like attributes. The robotic partner used in this experiment is a motor that generates a force at the end of the crank and is composed of two parts: a force based on a recording of the individual trials and a simulated inertia. The first part mimics the motions of a partner (Sect. 7.3.1) who has taken on the role of accelerating the crank by modifying the participant's own force trajectory that was recorded and averaged during the individual trials. A recorded version of the individual's forces was used so that variations in forces and completion times among participants would not affect the results.

The participants working with a robotic partner were provided with the forces similar to those found in the dyadic interactions. Each participant was assumed to take on the role of a deceleration specialized partner, so they would be completely responsible for all the force during deceleration, whereas the participant was free to choose their force during the acceleration phase. The robotic partner could have been programmed to take on the role of deceleration specialist, but every individual has the ability to complete either role, so it should not have made a difference. Since the acceleration role was easier to program, the robotic partner took on this role and applied enough force to accelerate both the crank and the participant's arm. However, the participants did not develop the specialized roles as expected, even though they consciously believed they were working with a human partner as discussed in Sect. 7.2.2. In terms of a Turing test, the results were split between consciously believing (i.e., passing) and physically acting different (i.e., failing). The human's forces did not show a very noticeable change even though the participants consciously believed the robotic partner was human [22].

With a participant's own amplified force applied as a feedforward force, the participants were given an easy way to specialize. There was no motor force during deceleration, so the participant was required to apply all the force, whereas they could choose their force during the acceleration phase. Comparing the applied force from a participant working alone to a participant working with a motor shows that they did not significantly change their feedforward force during the acceleration phase. They tended to accelerate similarly and actually applied slightly more peak force, but switched to deceleration earlier than they did when working alone. It seems that the participants were pushing with a preprogrammed feedforward force similar to their previous trials alone, but began to correct it and slowed down the crank earlier.

When working individually, each participant can accurately predict the result of their action. A human partner is less predictable, so the result of a cooperative action is slightly uncertain. Sebanz et al. [6] show that people can develop a representation of the actions of a person nearby when working on a complementary action. Presumably, humans have the same ability for haptic interactions, which would enable two people working together to depend on their partner's actions to complete the complementary action of specialization. It is expected that a person would also learn to depend on the robotic partner to complete the complementary action of specialization, but this was not the case. Scheidt et al. [9] show that people can adapt to unpredictable forces within one trial and, since the robotic partner's forces are more predictable than a human's forces, it is surprising that the participants did not learn to work with a predictable robotic partner in the same way as they did when working with an unpredictable human partner.

Lifting and moving a large or awkward object cannot be done individually, so people either get help from another person or, in some cases, a robot. Takubo et al. [77] demonstrate a robotic assistance system to empower a single human to work with a single robot to handle unwieldy objects. Their control strategy consists of a dominant human who leads while the robot emulates a virtual non-holonomic constraint. Initially, they kept the device in the horizontal plane allowing only 2-D motion. They experimentally validated the usability of this method by showing that participants could transport an object to an arbitrary location and orientation with only a small amount of learning by using "skills similar to using a wheelbarrow."

Takubo et al. [77] also extended the method to 3-D space. Interacting with the robot in 3-D space requires a substantial amount of planning in order to place it in an arbitrary location. In order to move the object directly down, the participant must first lower his end of the object and pull the object down along the declined angle to get the robot end at the correct height. At this point, the participant must raise his end and push the object back to the original horizontal location. This same strategy must be completed for each desired direction of motion. Although this control was inspired by human cooperation behavior, they are not trying to simulate the human interaction. Two humans cooperating can perform the task with one person attempting to act as a non-holonomic constraint, but this is not the most natural strategy. There are other cooperative strategies available. Two people can jointly maneuver an object to a desired location in a much more direct path, but this has the complicating factor that both people must know the goal destination. Learning to implement other strategies for how two humans cooperation haptically could help in creating more intuitive communication between a robot and a human.

Rahman et al. [78, 79] studied humans physically interacting using a one DoF placement task. They characterized the humans as either being a master or a follower. A master controlled the position of the object. The follower tracked the motion of the master with impedance control. The impedance model of the follower robot, discussed by Rahman et al. [80], changed the stiffness and damping throughout the motion. The impedance was high in the beginning and was essentially zero after 0.4 seconds. They determined the impedance mode by analyzing the resistance an arm will apply when it is led through a given path. By using the follower characteristics of a human response found in their studies, Rahman et al. [81] implemented the same response in a robot. Their aim was to make the robot imitate the response of a human when interacting with another human. Throughout this series of experiments, they do not discuss physical communication between the two humans and they do not take into account how the completion strategy changes from working alone to working with a partner.

Another study examined how a robot could act like a human who lacks knowledge of the end goal. Corteville et al. [82] gave an example of a blindfolded person who attempts to help a partner complete a task. They discussed how the blindfolded helper would wait for a trigger that indicates when the motion has begun before assisting. Throughout the task, the helper does not know where the leader is going, but the helper will begin to guess based upon the motions of the leader. Using an estimate of the motion, the helper will join in on the motion. Corteville et al. programed a robotic assistant based upon the motion estimation of a blindfolded human helper. The robotic assistant assumes that the path will follow a minimum jerk trajectory. An admittance controller provides assistance once an estimate of the operator's path is obtained. Based on the input from the participant, the controller will provide a scaled force along the predicted path that will aid in reaching the target. This design has the potential to directly take human control and apply it to a robotic controller. However, the robot motion is not adaptable to unknown targets in its current form. They assume that the start and end points are known to the controller, which are necessary for calculating the robotic trajectory. They mention that a first step of building versatile robots consists of "building up experience with simple tasks."

Another cooperative feedback mode (see [83] and Chap. 13 of this book) uses the concept of a negotiated interface point (NIP), which connects to the virtual object through a spring and damper. The NIP is similarly controlled using two haptic interface points, one of which is connected to the user and the other is connected to a controlled user. The role of the controller was set to switch between equal control, role blending, or user dominant, depending on the input from the user. The experiments were based on a target acquisition task and were conducted in two-dimensional space. Their results indicate that the model is more personal and more human-like.

7.5 Conclusions and Future Directions

This chapter provides an overview of how two humans interact and how this interaction can be extended for use in human-robot interaction. Generally, two people are faster on cooperative tasks compared to individually, but it comes at the price of much more energy exerted. Some of this energy is exerted against the partner as can be seen in specialization, where the two members of the dyad take on different temporal aspects of the task and in dyadic-contraction, where two individuals continue to push against each other even after the mutually held object is inside the target. Although co-contraction in individuals assists in the ability to reject perturbations, the similar dyadic-contraction effect does not provide the same benefit in dyads. When interacting with another agent, the perception of where the interaction force originates is also a factor in determining the response. Externally applied forces are viewed differently than self-generated forces and forces applied by a known non-human agent are perceived differently than the same forces applied by what appears to be a human agent.

There is still much to understand regarding how two humans interact and how best to implement human-robot interaction. Many of the current technologies will enable researchers to probe deeper into the minds and perceptions of physical group interaction using analysis techniques, such as fMRI, and interfaces such as brain-computer interfaces (BCI). These technologies will possibly reveal how people think about interacting with other people as opposed to many of the current methods that observe the result of the interactions. Understanding how one internally exploits the interaction with a partner will likely reveal improved methods for human-robot interaction.

References

1. Schmidt, R.C., Carello, C., Turvey, M.T.: Phase transitions and critical fluctuations in the visual coordination of rhythmic movements between people. J. Exp. Psychol. Hum. Percept. Perform. **16**(2), 227–247 (1990)
2. Schmidt, R.C., Turvy, M.T.: Phase-entrainment dynamics of visually couple rhythmic movements. Biol. Cybern. **70**, 369–376 (1994)
3. Amazeen, P.G., Schmidt, R.C., Turvey, M.T.: Frequency detuning of the phase entrainment dynamics of visually coupled rhythmic movements. Biol. Cybern. **72**, 511–518 (1995)
4. Pellegrino, G., Fadiga, L., Fogassi, L., Gallese, V., Rizzolatti, G.: Motor facilitation during action observation: a magnetic stimulations study. J. Neurophysiol. **73**, 2608–2611 (1992)
5. Rizzolatti, G., Fogassi, L., Gallese, V.: Neurophysiological mechanisms underlying the understanding and imitation of action. Nat. Rev., Neurosci. **2**(9), 661–670 (2001)
6. Sebanz, N., Knoblich, G., Prinz, W.: Representing others' actions: just like one's own? Cognition **88**, 11–21 (2003)
7. Sallnäs, E., Zhai, S.: Collaboration meets Fitts' law: Passing virtual objects with and without haptic force feedback. In: Proceedings of INTERACT, IFIP Conference on Human-Computer Interaction, pp. 97–104 (2003)
8. Sebanz, N., Bekkering, H., Knoblich, G.: Joint action: bodies and minds moving together. Trends Cogn. Sci. **10**(2), 70–76 (2006)
9. Scheidt, R.A., Dingwell, J.B., Mussa-Ivaldi, F.A.: Learning to move amid uncertainty. J. Neurophysiol. **86** (2001)
10. Wegner, N., Zeaman, D.: Team and individual performance on a motor learning task. J. Gen. Psychol. **55**, 127–142 (1956)
11. Shaw, R.E., Kadar, E., Sim, M., Repperger, D.W.: The intentional spring: A strategy for modeling systems that learn to perform intentional acts. J. Motiv. Behav. **24**(1), 3–28 (2011)
12. Knoblich, G., Jordon, J.S.: Action coordination in groups and individuals: Learning anticipatory control. J. Exp. Psychol. Learn. Mem. Cogn. **29**(5), 1006–1016 (2003)
13. Moffett, M.W.: Cooperative food transport by an asiatic ant. Natl. Geogr. Res. **4**, 386–394 (1988)
14. Kube, C.R., Bonabeua, E.: Cooperative transport by ants and robots. Robot. Auton. Syst. **30**, 85–101 (2000)
15. Bonabeau, E., Dorigo, M., Theraulaz, G.: Inspiration for optimization from social insect behaviour. Nature **406**, 39–42 (2000)
16. Summers, L.G., Shannon, J.H., White, T.R., Shiner, R.J.: Fly-by-wire sidestick controller evaluation. SAE Technical Paper 871761, Aerospace Technology Conference and Exposition, Long Beach, CA, Society of Automotive Engineers (1987)
17. Field, E., Harris, D.: A comparative survey of the utility of cross-cockpit linkages and autoflight systems' backfeed to the control inceptors of commercial aircraft. Ergonomics **41**(10), 1462–1477 (1998)
18. Reed, K.B., Peshkin, M.A., Hartmann, M.J., Grabowecky, M., Patton, J., Vishton, P.M.: Haptically linked dyads: Are two motor-control systems better than one? Psychol. Sci. **17**(5), 365–366 (2006)
19. Reed, K.B., Peshkin, M., Hartmann, M.J., Patton, J., Vishton, P.M., Grabowecky, M.: Haptic cooperation between people, and between people and machines. In: Proc. IEEE/RSJ Int. Conf. Intell. Robots Syst., pp. 2109–2114 (2006)
20. Glynn, S., Henning, R.: Can teams outperform individuals in a simulated dynamic control task. In: Proceedings of the Human Factors and Ergonomics Society Annual Meeting, vol. 6, pp. 141–144 (2000) (ISSN: 10711813)
21. Glynn, S., Fekieta, R., Henning, R.: Use of force-feedback joysticks to promote teamwork in virtual teleoperation. In: Virtual Teleoperation Proc. of the Human Factors and Ergonomics Society 45th Annual Meeting (2001)
22. Reed, K.B., Peshkin, M.A.: Physical collaboration of human-human and human-robot teams. IEEE Trans. Haptics **1**(2), 108–120 (2008)

23. Reed, K.B., Patton, J., Peshkin, M.: Replicating human-human physical interaction. In: Proc. IEEE Int. Conf. Robot. Autom., pp. 3615–3620 (2007)
24. Turing, A.: Computing machinery and intelligence. Mind **LIX**(236), 433–460 (1950)
25. Steuer, J.: Defining virtual reality: Dimensions determining telepresence. J. Commun. **42**(4), 73–93 (1992)
26. Canny, J., Paulos, E.: Tele-embodiment and shattered presence: Reconstructing the body for online interaction. In: The Robot in the Garden: Telerobotics and Telepistemology in the Age of the Internet, pp. 276–294 (2000)
27. Zajonc, R.B.: Social facilitation. Science **149**, 269–274 (1965)
28. Triplett, N.: The dynamogenic factors in pacemaking and competition. Am. J. Psychol. **9**, 507–533 (1898)
29. Schmitt, B.H., Gilovich, T., Goore, N., Joseph, L.: Mere presence and social facilitation: One more time. J. Exp. Soc. Psychol. **22**, 242–248 (1986)
30. Fitts, P.M.: The information capacity of the human motor system in controlling the amplitude of movement. J. Exp. Psychol. **47**, 381–391 (1954)
31. Mottet, D., Guiard, Y., Ferrand, T., Bootsma, R.J.: Two-handed performance of a rhythmical Fitts task by individuals and dyads. J. Exp. Psychol. Hum. Percept. Perform. **27**(6), 1275–1286 (2001)
32. Reed, K.B., Peshkin, M., Colgate, J.E., Patton, J.: Initial studies in human-robot-human interaction: Fitts' law for two people. In: Proc. IEEE Int. Conf. Robot. Autom., pp. 2333–2338 (2004)
33. Accot, J., Zhai, S.: Beyond Fitts' law: Models for trajectory-based HCI tasks. In: Proceedings of the CHI, vol. 97, pp. 295–302 (1997)
34. Looser, J.C.A.: 3d games as motivation in Fitts' law experiments. Masters thesis, University of Canterbury (2002)
35. Hinckley, K., Cutrell, E., Bathiche, S., Muss, T.: Quantitative analysis of scrolling techniques. In: Conference on Human Factors in Computer Systems, pp. 65–72 (2002)
36. Zhai, S., Conversy, S., Beaudouin-Lafon, M., Guiard, Y.: Human on-line response to target expansion. In: Proceedings of CHI 2003, ACM Conference on Human Factors in Computing Systems, Fort Lauderdale, Florida, pp. 177–184 (2003)
37. Shergill, S.S., Bays, P.M., Frith, C.D., Wolpert, D.M.: Two eyes for an eye: The neuroscience of force escalation. Science **301**, 187 (2003)
38. Christian, W.: Multiple humans interacting with a robot to obtain the fundamental properties of materials. Master's thesis, University of South Florida (2010)
39. Gentry, S., Feron, E., Murray-Smith, R.: Human-human haptic collaboration in cyclical Fitts' tasks. In: Proc. IEEE/RSJ Int. Conf. Intell. Robots Syst (2005)
40. Gentry, S.: Dancing cheek to cheek: Haptic communication between partner dancers and swing as a finite state machine. Ph.D. thesis, Massachusetts Institute of Technology (2005)
41. Holldampf, J., Peer, A., Buss, M.: Synthesis of an interactive haptic dancing partner. In: Proc. IEEE RO-MAN, pp. 527–532 (2010)
42. Karniel, A., Meir, R., Inbar, G.F.: Exploiting the virtue of redundancy. In: International Joint Conference on Neural Networks, Washington, DC (1999)
43. Lacquaniti, F., Maioli, C.: Distributed control of limb position and force. In: Stelmach, G.E., Requin, J. (eds.) Tutorials in Motor Behavior II, pp. 31–54 (1992)
44. Reed, K.B., Peshkin, M., Hartmann, M.J., Colgate, J.E., Patton, J.: Kinesthetic interaction. In: Proc. IEEE Int. Conf. Rehabilitation Robotics, pp. 569–574 (2005)
45. Groten, R., Feth, D., Goshy, H., Peer, A., Kenny, D.A., Buss, M..: Experimental analysis of dominance in haptic collaboration. In: Proc. IEEE RO-MAN, pp. 723–729 (2009)
46. Flash, T., Hogan, N.: The coordination of arm movements: An experimentally confirmed model. J. Neurosci. **5**(7), 1688–1703 (1985)
47. Feygin, D., Keehner, M., Tendick, F.: Haptic guidance: Experimental evaluation of a haptic training method for a perceptual motor skill. In: Proceedings of the 10th Symposium on Haptic Interfaces for Virtual Environments and Teleoperator Systems (HAPTICS) (2002)
48. Rosenbaum, D.A.: Human movement initiation: Specification of arm, direction, and extent. J. Exp. Psychol. Gen. **109**, 444–474 (1980)

49. Hallett, M., Shahani, B., Young, R.: EMG analysis of stereotyped voluntary movements in man. J. Neurol. Neurosurg. Psychiatry **38**, 1154–1162 (1975)
50. Hannaford, B., Stark, L.: Roles of the elements of the triphasic control signal. Exp. Neurol. **90**, 619–634 (1985)
51. Gottlieb, G.L., Corcos, D.M., Agarwal, G.C., Latash, M.L.: Principles underlying single joint movement strategies. In: Winters, J.M., Woo, S.L.Y. (eds.) Multiple Muscle Systems, pp. 236–250. Springer, New York (1990)
52. Gottlieb, G.L., Corcos, D.M., Agarwal, G.C.: Organizing principles for single-joint movements. I. A speed insensitive strategy. J. Neurophysiol. **62**, 342–357 (1989)
53. Gottlieb, G.L., Chen, C.H., Corcos, D.M.: Nonlinear control of movement distance at the human elbow. Exp. Brain Res. **112**, 289–297 (1996)
54. Ramos, C.F., Hacisalihzade, S.S., Stark, L.W.: Behavior space of a stretch reflex model and its implications for the neural control of voluntary movement. Med. Biol. Eng. Comput. **28**, 15–23 (1990)
55. Prodoehl, J., Gottlieb, G., Corcos, D.: The neural control of single degree-of-freedom elbow movements: Effect of starting joint position. Exp. Brain Res. **153**, 7–15 (2003)
56. Pan, P., Peshkin, M.A., Colgate, J.E., Lynch, K.M.: Static single-arm force generation with kinematic constraints. J. Neurophysiol. **93**, 2752–2765 (2005)
57. Feth, D., Groten, R., Peer, A., Hirche, S., Buss, M.: Performance related energy exchange in haptic human-human interaction in a shared virtual object manipulation task. In: Proc. and Symposium on Haptic Interfaces for Virtual Environment and Teleoperator Systems. Third Joint EuroHaptics Conference. World Haptics 2009, 8–20 March 2009, pp. 338–343 (2009)
58. Pham, H.T.T., Ueha, R., Hirai, H., Miyazaki, F.: A study on dynamical role division in a crank-rotation task from the viewpoint of kinetics and muscle activity analysis. In: Proc. IEEE/RSJ Int Intelligent Robots and Systems (IROS) Conf., pp. 2188–2193 (2010)
59. Ueha, R., Pham, H.T.T., Hirai, H., Miyazaki, F.: A simple control design for human-robot coordination based on the knowledge of dynamical role division. In: Proc. IEEE/RSJ Int. Conf. Intelligent Robots and Systems IROS 2009, pp. 3051–3056 (2009)
60. Ueha, R., Pham, H.T.T., Hirai, H., Miyazaki, F.: Dynamical role division between two subjects in a crank-rotation task. In: Proc. IEEE Int. Conf. Rehabilitation Robotics ICORR 2009, pp. 701–706 (2009)
61. Reinkensmeyer, D.J., Lum, P.S., Lehman, S.L.: Human control of a simple two-hand grasp. Biol. Cybern. **67**(6), 553–564 (1992)
62. Swinnen, S.P., Wenderoth, N.: Two hands, one brain: cognitive neuroscience of bimanual skill. Trends Cogn. Sci. **8**(1), 18–25 (2004)
63. Malabet, H.G., Robles, R.A., Reed, K.B.: Symmetric motions for bimanual rehabilitation. In: Proc. IEEE/RSJ Int Intelligent Robots and Systems (IROS) Conf., pp. 5133–5138 (2010)
64. McAmis, S., Reed, K.B.: Symmetry modes and stiffnesses for bimanual rehabilitation. In: Proc. of the 12th Intl. Conf. on Rehabilitation Robotics (ICORR), pp. 1106–1111 (2011)
65. Osu, R., Franklin, D.W., Kato, H., Gomi, H., Domen, K., Yoshioka, T., Kawato, M.: Short- and long-term changes in joint co-contraction associated with motor learning as revealed from surface EMG. J. Neurophysiol. **88**, 991–1004 (2001)
66. Milner, T.E.: Adaptation to destabilizing dynamics by means of muscle cocontraction. Exp. Brain Res. **143**, 406–416 (2002)
67. Franklin, D.W., Burdet, E., Osu, R., Kawato, M., Milner, T.E.: Functional significance of stiffness in adaptation of multijoint arm movements to stable and unstable dynamics. Exp. Brain Res. **151**, 145–157 (2003)
68. Shadmehr, R., Mussa-lvaldi, F.A.: Adaptive representation of dynamics during learning of a motor task. J. Neurosci. **14**(5), 3208–3224 (1994)
69. Patton, J.L., Elkins, P.: Training with a Bimanual-Grasp Beneficially Influences Single Limb Performance. Society for Neuroscience, Orlando (2002)
70. Chib, V.S., Patton, J.L., Lynch, K.M., Mussa-Ivaldi, F.A.: Haptic discrimination of perturbing fields and object boundaries. HAPTICS 2004 0-7695-2112-6/04 (2004)
71. Perreault, E.J., Kirsch, R.F., Crago, P.E.: Effects of voluntary force generation on the elastic components of endpoint stiffness. Exp. Brain Res. **141**, 312–323 (2001)

72. Gomi, H., Osu, R.: Task-dependent viscoelasticity of human multijoint arm and its spatial characteristics for interaction with environments. J. Neurosci. **18**(21), 8965–8978 (1998)
73. Choi, S., Walker, L., Tan, H.Z., Crittenden, S., Reifenberger, R.: Force constancy and its effect on haptic perception of virtual surfaces. ACM Trans. Appl. Percept. **2**(2), 89–105 (2005)
74. Chib, V.S., Patton, J.L., Lynch, K.M., Mussa-Ivaldi, F.A.: Haptic identification of surfaces as fields of force. J. Neurophysiol. **95**, 1068–1077 (2006)
75. Reed, K.B.: Understanding the haptic interactions of working together. Ph.D. thesis, Northwestern University (2007)
76. Akazawa, K., Milner, T.E., Stein, R.B.: Modulation of reflex EMG and stiffness in response to stretch of human finger muscle. J. Neurophysiol. **49**(1), 16–27 (1983)
77. Takubo, T., Arai, H., Hayashibara, Y., Tanie, K.: Human-robot cooperative manipulation using a virtual nonholonomic constraint. Int. J. Robot. Res. **21**, 541–553 (2002)
78. Rahman, M., Ikeura, R., Mizutani, K.: Cooperation characteristics of two humans in moving an object. Mach. Intell. Robot. Control **4**(2), 43–48 (2002)
79. Rahman, M., Ikeura, R., Mizutani, K.: Analysis of cooperation characteristics of two humans in moving an object. In: Proceedings of the International Conference on Mechatronics and Information Technology, vol. 19, pp. 454–458 (2001)
80. Rahman, M., Ikeura, R., Mitzutani, K.: Impedance characteristics of human arm for cooperative robot. In: International Conference on Control, Automation and Systems, pp. 1455–1460 (2002)
81. Rahman, M., Ikeura, R., Mizutani, K.: Control characteristics of two humans in cooperative task and its application to robot control. In: Proceedings of 2000 International Conference on Industrial Electronics Control and Instrumentation, pp. 1773–1778 (2000)
82. Corteville, B., Aertbelien, E., Bruyninckx, H., Schutter, J.D., Brussel, H.V.: Human-inspired robot assistant for fast point-to-point movements. In: Proc. IEEE Int. Conf. Robot. Autom. (2007)
83. Oguz, S.O., Kucukyilmaz, A., Sezgin, T.M., Basdogan, C.: Haptic negotiation and role exchange for collaboration in virtual environments. In: Proc. IEEE Haptics Symp., pp. 371–378 (2010)

Part II
Technology and Rendering

Chapter 8
Data-Driven Visuo-Haptic Rendering of Deformable Bodies

Matthias Harders, Raphael Hoever, Serge Pfeifer, and Thibaut Weise

Abstract Our current research focuses on the investigation of new algorithmic paradigms for the data-driven generation of sensory feedback. The key notion is the collection of all relevant data characterizing an object as well as the interaction during a recording stage via multimodal sensing suites. The recorded data are then processed in order to convert the raw signals into abstract descriptors. This abstraction then also enables us to provide feedback for interaction which has not been observed before. We have developed a first integrated prototype implementation of the envisioned data-driven visuo-haptic acquisition and rendering system. It allows users to acquire the geometry and appearance of an object. In this chapter we outline the individual components and provide details on necessary extensions to also accommodate interaction scenarios involving deformable objects.

8.1 Introduction

A focus area of current research in multi-modal simulation is data-driven recording and rendering of objects, and their respective manipulations. The key notion is to visually, as well as haptically capture the interaction of a user with an object during unconstrained manipulation via sensing systems. Recorded interaction data are analyzed and prepared for the subsequent rendering phase. The target of the data analysis step is to extract key patterns characterizing the object, thus allowing to provide feedback, which has not been observed during the recording. This avoids explicit model-based computations; thus data-driven rendering techniques provide an alternative to approaches based on physical or heuristic models. The strength of the former is the ability to display complex phenomena with high accuracy, while avoiding complex models.

In previous work we have developed various components required for data-driven visuo-haptic rendering. In [1–3] a framework for providing haptic feedback is presented. Further, we also proposed a system for real-time visual acquisition of objects in [4, 5]. While the haptic component focused on deformable objects, the visual

M. Harders (✉) · R. Hoever · S. Pfeifer · T. Weise
Computer Vision Lab, ETH Zurich, Sternwartstrasse 7, 8092 Zurich, Switzerland
e-mail: mharders@vision.ee.ethz.ch

A. Peer, C.D. Giachritsis (eds.), *Immersive Multimodal Interactive Presence*,
Springer Series on Touch and Haptic Systems,
DOI 10.1007/978-1-4471-2754-3_8, © Springer-Verlag London Limited 2012

framework mainly targeted rigid entities. In recent work we focused on the combination of these system to obtain a framework for data-driven visuo-haptic rendering of deformable bodies. In this chapter we will outline the different steps to merge the two frameworks, as well as the extensions required to deal with deformable objects.

In the next section, we provide an overview of related work. First previous activities in the fields of real-time 3D visual scanning and data-driven haptic rendering will be scrutinized. Thereafter, the system components for visual and haptic acquisition and rendering will be introduced. This is followed by details on the extensions to deformable objects. Next, an overview of a prototype demonstrator system will be given. Finally, the chapter concludes with a summary of the discussed work.

8.2 Related Work

8.2.1 Real-Time 3D Scanning

Many different non-contact shape scanning systems have been proposed in the past. In general these methods can be divided into active and passive systems. Passive systems rely only on the images as captured by one or more cameras to reconstruct the 3D shape of a scene. Most work in this area has been done based on stereo (two cameras) setups. Many different algorithms for stereo reconstruction have been developed. A comprehensive overview and evaluation can be found in [6]. The same authors also provide a benchmark on the web [7], which allows developers to test and compare algorithms. A recent top performer [8] relies on color segmentation and optimization on the segment level. However, the major problem for stereo systems is that they cannot handle uniformly colored regions well due to the intrinsic matching ambiguity between the two images.

Active systems project light into the scene and analyze the recorded image. Several different techniques exist. A good overview can be found in [9]. The major techniques are the laser scanner, the line scanner, the time-of-flight system, and the pattern projection system.

Laser scanners project a single laser point into the scene. The depth is either reconstructed using triangulation with the detected image in a CCD line camera or using the time-of-flight of the laser signal. A line scanner projects a laser line and uses a CCD array to detect the line in the scene. It allows more reconstructions per second than a point-based scanner, though at lower accuracy. However, both point and line-based scanners do not provide dense reconstructions. This means that they are not well-suited for handling dynamic scenes.

Time-of-Flight systems [10] project a pulse or a phase-modulated light and detect the signal in a CCD or CMOS array. The time for the light to travel to the object and back to the sensor is measured and used to calculate the distance. Each cell of the sensor calculates the depth and (gray-scale) intensity resulting in a real-time shape acquisition system. However, high cost and low resolution are (still) the major drawback.

Pattern projection systems project a known pattern into the scene and use software to decode the pattern [11, 12]. The projector can be seen as an inverse camera. Triangulation is performed between camera and projector. The major difficulty is the decoding of the pattern in the image. Binary gray-code patterns encode each projector pixel in a series of images which can be directly calculated from the recorded intensities [13]. Accuracy is increased by using a sinusoidal phase shifted pattern and calculating the phase directly [13, 14].

The authors of [15] and [16] present recent systems combining structured light and stereo into a common framework.

One-shot techniques only require a single image for the 3D reconstruction and are particularly suitable for dynamic scenes. For example [17] and [18] project a series of lines which are detected in the image. The major difficulty here is the correct labeling of the detected lines.

8.2.2 Data-Driven Haptic Rendering

Force signals recorded from real objects are used in haptic research in a wide variety of approaches. These can be classified by the strategy how to incorporate the information from the recordings into a rendering algorithm.

One data utilization method is the extraction of parameterized behavior models from measured force signals. For instance, in [19], parameters like stiffness, damping, and inertia were obtained by fitting a piecewise linear model to recorded force signals. In another project, cutting forces from surgical scissors were first acquired for different tissues [20]. These signals were then used for haptic feedback using tuned, piecewise linear models [21]. In a similar manner the recorded force signals from a pin insertion task were modeled empirically by a polynomial function in [22]. Further work focusing on force profiles of push-buttons was described in [23, 24]. A non-linear model was fitted to the captured force and acceleration signals. The obtained model parameters were then used to provide force feedback. A system to capture human motion along with the contact forces that occur during the interaction with rigid objects was presented in [25]. The described algorithm allows one to estimate the compliance of the finger joints. A limitation of those studies was the a-priori defined model, which included somewhat simplified assumptions about the material. Errors might also have occurred due to the fitting steps. These drawbacks can be avoided in our approach.

Other related work examined the acquisition of specific surface properties, such as texture or friction. In [26], surface properties like texture and compliance were obtained using the hand-held device developed in [27]. A heuristic model to estimate compliance was also used, however, the accuracy was not discussed. In [28] a method to estimate the parameters of a friction model together with the inertia of the object was proposed.

Additional work examined the enhancement of model-based rendering using data obtained from recorded interactions. In [29, 30], an event-based approach for haptic replay was proposed, focusing on tapping on hard objects. The method captured acceleration profiles, from which a library of force transients was derived. The latter could be used to enhance conventional haptic rendering.

Some previous research examined force superposition assuming linear models. A related algorithm was proposed in [31, 32], which was based on superposition of force signals prerecorded with a fully automatic robotic measurement facility. Due to the assumed linearity, the algorithm mainly provided appropriate results for small deformations. Linear visco-elastic tissue was considered in [33, 34] in a precalculation-based algorithm for which parameters were obtained from force recordings using real tissue. Besides the pure viscous damping this work also addressed stress relaxation.

In [35, 36] a method was presented how to interpolate distinct force fields from different contact points over the whole object surface. In order to generate contact forces of a non-linear material, one-dimensional force deflection curves were precalculated based on measurements and used for force feedback generation. While this approach could capture non-linearity of a material, it did not address visco-elasticity.

8.3 System Components

8.3.1 Visual Recording

For the visual acquisition of object geometry and texture we have developed a real-time 3D in-hand scanning system [4]. It is based on a phase-shift acquisition paradigm. It uses a combination of sinusoidal phase-shift structured light and stereo matching, thus offering the benefits of both established methods. Maximum accuracy is achieved through phase-shift, while stereo information provides robustness to discontinuities.

The hardware setup consists of a standard DLP projector (InFocus IM38), two high-speed monochrome cameras (Allied Vision Tec PIKE F-032B), a color camera (Allied Vision Tec STINGRAY F-046C), and a micro-controller for synchronization (Microchip PICDEM HPC Explorer Board). The color wheel of the projector has been removed, so that it projects three independent monochrome images at 120 Hz which are sent to the projector as the red, green, and blue color channel. The two monochrome cameras are synchronized and record the three images. The delay between two images is currently 4 ms, where each exposure takes 2 ms, resulting in a total recording time of 14 ms for the three phase-shifted images. The texture camera is also synchronized, but uses a longer exposure to integrate over all three projected images. Figure 8.1 depicts the visual scanning setup.

Phase-shift is a well-known fringe projection method for the retrieval of 3D shape information about objects. A set of phase-shifted sinusoidal patterns are projected and the phase is calculated at each point. The minimum number of images is three, but more images improve the accuracy of the reconstructed phase. However, we use only three images as it allows high speed projection using the modified DLP.

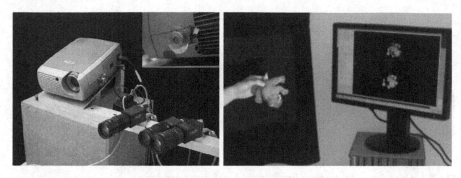

Fig. 8.1 Overview of visual recording setup and real-time acquisition

The intensities for each pixel (x_i, y_i) of the three images can be described by the following formulas assuming a linear projector, a linear camera, constant lighting, and a static object during the recording interval (note that (x_i, y_i) is omitted for the sake of brevity):

$$I_r = I_{dc} + I_{mod} \cdot \cos(\phi - \theta),$$
$$I_g = I_{dc} + I_{mod} \cdot \cos(\phi), \qquad\qquad (8.1)$$
$$I_b = I_{dc} + I_{mod} \cdot \cos(\phi + \theta),$$

where I_r, I_g, and I_b are the recorded intensities, I_{dc} is the DC component, I_{mod} is half the signal amplitude, ϕ is the phase and θ the constant phase-shift. The wrapped phase can be calculated as

$$\phi' = \arctan\left(\tan\left(\frac{\theta}{2}\right) \frac{(I_r - I_b)}{(2I_g - I_r - I_b)} \right). \qquad (8.2)$$

Next, two-dimensional phase unwrapping is carried out to convert the wrapped to the absolute phase via:

$$\phi(x, y) = \phi'(x, y) + 2\pi \cdot k(x, y), \quad k(x, y) \in [0, N - 1] \qquad (8.3)$$

where $k(x, y)$ represents the period and N is the number of projected periods. Different methods for unwrapping have been proposed in the past. In general these unwrap the phase by integrating along reliable paths in the image or by taking a global minimum-norm approach. Unfortunately, these methods only provide a relative unwrapping and do not solve for the absolute phase. Instead, we rely on stereo matching between two cameras to resolve the period.

For each possible period k, the 3D position is calculated using point-surface triangulation between the first camera and the projector. The resulting 3D point is projected into the second camera, and the correlation value between the two camera pixels is calculated as a sum-of-squared differences. The period with maximum correlation (minimum SSD) is assumed to be the true period of the recorded phase. In

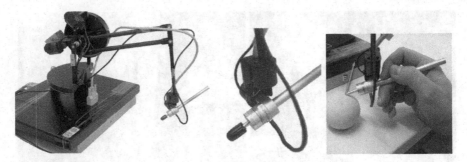

Fig. 8.2 Overview of haptic recording setup and real-time acquisition

order to improve reconstruction accuracy, left-right consistency checks are carried out to detect outliers. The checks ensure that every assigned absolute phase value is consistent between both cameras.

Real-time performance is achieved by implementing most of the algorithms on a GPU in form of vertex and fragment shaders. A continuous frame rate of 17 frames per second on an NVidia GeForce 7900 GTX resulted. The reconstruction accuracy of the system is high with only 0.125 mm error at 1100 mm distance.

8.3.2 Haptic Recording

For the haptic component also a special recording system has been developed [1]. With the targeted data-driven synthesis strategy in mind, we use the same setup for recording and rendering. The overall system is depicted in Fig. 8.2.

The setup consists of a PHANToM Premium 1.5 from SensAble which provides encoder-based position measurements with an accuracy of 0.03 mm in the center of the workspace. A custom-designed probing tool is attached to the device, which is equipped with several sensors. A 6DOF force sensor (ATI Nano17) is used to measure contact forces. An acceleration sensor (ADXL330, Analog Devices) is employed to provide accurate measurements of acceleration. Based on their readings the tool dynamics are determined. Furthermore, an optical sensor has also been integrated into a different version of the tool to provide slip measurements. However, this prototype will not be further considered here.

In order to obtain correctly registered force and position signals, we have developed a specialized real-time environment. A RTAI machine with real-time drivers has been equipped with different boards for high-fidelity data acquisition. The position signals of the PHANToM encoders are captured via a Sensoray DAQ board (Model 626) and the force signals of the ATI sensor via a National Instruments PCI-6220 board. This approach allows us to obtain both signals synchronously with very low jitter. In addition, the setup permits using sampling frequencies up to 5 kHz. The real-time RTAI system runs on a Pentium IV 3 GHz machine.

In order to generate data-driven haptic feedback we first record the feedback when interacting with real objects using the probing tool. The device records the trajectory along which the user moves the tool. Using the position encoders and an additional acceleration sensor, the position, velocity, and acceleration of the tool are estimated by a Kalman filter. Also, the contact forces at the tool tip are recorded. These raw data are then interpolated. To this end, the independent variables of the interpolation function need to be defined beforehand. For instance, to interpolate the elastic properties of the material, the recorded force needs to be interpolated over the tool position. If also dynamic effects need to be considered, additional input dimensions—like the tool velocity or acceleration—need to be incorporated. For the visco-elastic materials, we use the tool position, velocity, and a low-pass filtered version of the tool velocity. The latter is important to capture transient material effects like force relaxation. We employed a first order low-pass filter given by

$$\mathbf{v}_{out}[nT] = a \cdot \mathbf{v}_{out}[(n-1)T] + (1-a) \cdot \mathbf{v}_{in}[nT] \quad \text{with } a = \exp\left[-\frac{T}{\tau}\right], \quad (8.4)$$

where T is the sample period and τ is the time constant of the filter. For a one-dimensional tool interaction, it follows that the input space of the interpolator can be represented by a three dimensional interaction vector $\mathbf{x}$. For the generic interpolation, a sum of radial basis functions (RBF) is fit to the data. This sum can be expressed as

$$h(\mathbf{x}) = \sum_{j=1}^{M} w_j \phi(\|\mathbf{x} - \mathbf{c}_j\|) + \sum_{k=1}^{L} d_k g_k(\mathbf{x}) \mathbf{x} \in \mathbb{R}^n, \quad (8.5)$$

with the radial basis functions ϕ, the RBF centers $\mathbf{c}_j$, and the linear weights w_j. The functions g_k form a basis of an additional polynomial term with the coefficients d_k.

8.4 Extension to Deformable Objects

8.4.1 Visual Capture of Deformations

Overview

Capturing the geometry of an object together with its deformations while being manipulated is a highly under-constrained problem. We therefore opt for a two-step strategy. First, a user turns an object in front of the 3D scanner. Using our online model building procedure [5] a 3D mesh of the object is automatically reconstructed. Thereafter, the user interacts with and deforms the object, while the 3D scanner captures any surface deformations. These deformations are extracted as vertex displacements using template-based non-rigid registration. The resulting displacement vectors are finally used for deformation interpolation as described in Sect. 8.4.2. In the following, the steps of the non-rigid registration procedure for capturing deformations are described in more detail.

Non-rigid Registration

The target of this registration process is to deform an initial 3D mesh in such a way that it fits to a currently scanned geometry. Non-rigid registration methods typically formulate deformable registration as an optimization problem consisting of a mesh smoothness term and several data fitting terms (see e.g. [37–39]).

We represent deformations with displacement vectors $\mathbf{d}_i = \tilde{\mathbf{v}}_i - \mathbf{v}_i$ for each mesh vertex $\mathbf{v}_i \in \mathcal{V}$ and deformed mesh vertex $\tilde{\mathbf{v}}_i \in \tilde{\mathcal{V}}$. Additionally, the deformed mesh may undergo a rigid transformation $[\mathbf{R}, \mathbf{t}]$:

$$\hat{\mathbf{v}}_i = \mathbf{R}(\mathbf{v}_i - \mathbf{d}_i) + \mathbf{t}. \tag{8.6}$$

The unknowns in this non-linear equation are the rigid transformation and the vertex displacements. In order to keep the system linear, the registration is usually split into a rigid and a non-rigid part.

First, the mesh is registered rigidly with the input geometry using the Iterative Closest Point (ICP) algorithm with the given fixed displacements. The displacements are then optimized in a second non-rigid registration step in which the rigid transformation remains fixed. This two-step procedure is iterated until the error drops below a threshold.

For the non-rigid registration, an energy term is minimized which is based on a globally elastic deformation model and geometric/image data terms. The former terms target deformation smoothness by minimizing stretch and/or bending of the underlying deformation field. Stretch is reduced by minimizing the first order derivatives of the displacement vectors (approximated following [40]) via the energy term:

$$E_{stretch} = \sum_{i \in \mathcal{V}} \sum_{j \in \Omega(i)} \frac{\|\mathbf{d}_i - \mathbf{d}_j\|_2^2}{\|\mathbf{v}_i - \mathbf{v}_j\|_2^2}. \tag{8.7}$$

This expression minimizes the displacement vector differences of neighboring template vertices, scaled by the distance of the vertices in the undeformed template mesh. However, sparse normal displacement constraints in combination with stretching energy lead to strong local deformations.

Bending of the deformation field is reduced by minimizing the second order derivatives of the displacement vectors, using the standard cotangent discretization of the Laplace-Beltrami operator (see [41]), via the energy term:

$$E_{bend} = \sum_{i \in \mathcal{V}} \|\delta \mathbf{d}_i\|_2^2. \tag{8.8}$$

Deformations are distributed more nicely with this formulation due to C1 smoothness. It should be noted that the global rigid transformation for registration is essential for both terms. The elastic deformation model is a linear approximation of a non-linear shell deformation and thus cannot handle large rotations accurately.

The elastic terms are suitable for dealing with a wide range of deformations while still enabling efficient processing of extended visual acquisition sequences. In addition to the elastic terms, geometric and image terms are included in the metric.

The distance of each mesh vertex $\hat{\mathbf{v}}_i$ to some 3D point correspondence $\mathbf{c}_i$ is minimized in a point-to-point geometric 3D constraint:

$$E_{p2p} = \sum_{i \in \mathcal{V}} \|w_i^{p2p} \hat{\mathbf{v}}_i - \mathbf{c}_i\|_2^2, \tag{8.9}$$

with w_i^{p2p} being a correspondence weighting term. A similar point-to-plane metric minimizes the distance of each mesh vertex $\hat{\mathbf{v}}_i$ to the tangential surface defined at some 3D point correspondence $\mathbf{c}_i$ with surface normal $\mathbf{n}_i$:

$$E_{p2s} = \sum_{i \in \mathcal{V}} \|w_i^{p2s} \mathbf{n}_i^T (\hat{\mathbf{v}}_i - \mathbf{c}_i)\|_2^2. \tag{8.10}$$

Again, w_i^{p2s} is a weighting term. Note that the point-to-point constraint is equivalent to adding three individual point-to-plane constraints with each coordinate axis as normal. The final energy term is a 2D image space constraint. Similar to the geometric point-to-plane constraint, the 2D point-to-texture constraint minimizes the distance of the projection $\Pi(\hat{\mathbf{v}}_i)$ of vertex $\hat{\mathbf{v}}_i$ to some 2D line defined by a 2D normal $\mathbf{n}_i^{tex}$ and 2D point $\mathbf{c}_i^{tex}$ in the image space of a given camera:

$$E_{tex} = \sum_{i \in \mathcal{V}} \|w_i^{tex} \mathbf{n}_i^{tex} (\Pi(\hat{\mathbf{v}}_i) - \mathbf{c}_i^{tex})\|_2^2. \tag{8.11}$$

In order to ensure linearity, a weak perspective camera model is used for the projection:

$$\Pi(\mathbf{x}_i) = \frac{f}{\bar{z}_i} \begin{bmatrix} 1 & 0 & 0 \\ 0 & 1 & 0 \end{bmatrix} (\mathbf{R}_{cam} \mathbf{x}_i + \mathbf{t}_{cam}), \tag{8.12}$$

where $\bar{z}_i$ is the fixed depth of the current mesh vertex with respect to the camera, f the focal length, and $(\mathbf{R}_{cam}, \mathbf{t}_{cam})$ the extrinsic camera parameters. Note that optical flow is a special case of this generic image space constraint. Also, texture point correspondences may be added as two point-to-texture constraints with the image coordinate axes as normals.

Both the smoothness and data terms are linear in the parameters. Thus, the optimal non-rigid deformation is found as a linear least squares solution of:

$$\underset{\mathbf{d}_i \in \mathcal{V}}{\operatorname{argmin}} \, \alpha_s E_{stretch} + \alpha_b E_{bend} + \alpha_{p2p} E_{p2p} + \alpha_{p2s} E_{p2s} + \alpha_{tex} E_{tex}. \tag{8.13}$$

We typically minimize either stretch ($\alpha_s = 1$) or bending ($\alpha_b = 1$), but both energies may also be combined. The resulting over-determined linear system is sparse and can be solved efficiently via Cholesky decomposition [42]. The proposed method may be used for any template-based non-rigid registration. In the following we describe how this technique is used for capturing the deformation of an object manipulated by the user.

Deformation Tracking

The non-rigid registration framework introduced above is used to track the deformations of an object during manipulation. For this purpose we use both geometric and texture constraints. The 3D scanner captures the geometry and texture of a scene with high temporal frequency. Given that the rigid transformation and the deformation vectors of the previous step are known, we can simply use projective closest-point correspondences for the geometric point-to-point and point-to-plane constraints, as well as optical flow as point-to-texture constraints.

We combine point-to-plane minimization (E_{p2s}) with a small point-to-point regularization (E_{p2p}) as described in [43] to prevent oscillations:

$$E_{fit} = \sum_{i \in \mathcal{V}} w_i^{fit} (\|\mathbf{n}_{\mathbf{c}_i}^T (\hat{\mathbf{v}}_i - \mathbf{c}_i)\|_2^2 + 0.1\|\hat{\mathbf{v}}_i - \mathbf{c}_i\|_2^2). \qquad (8.14)$$

Closest point pairs with incompatible normals or distance larger than 10 mm are pruned by setting their corresponding weights to $w_i^{fit} = 0$.

Optical flow [44] is used to enhance template tracking by establishing interframe correspondences from video data. Instead of using an independent optical flow procedure as in [40], we directly include the optical flow constraints into the optimization, similar to model-based tracking methods [45]. We thus avoid solving the difficult 2D optical flow problem and integrate the constraints directly into the 3D optimization:

$$E_{opt} = \sum_{i \in \mathcal{V}} w_i^{opt} (\nabla g_{t,i} \mathbf{\Pi}(\hat{\mathbf{v}}_i^{t+1} - \hat{\mathbf{v}}_i^t) + g_{t+1,i} - g_{t,i}), \qquad (8.15)$$

where $g_{t,i} = g_t(\mathbf{\Pi}(\tilde{\mathbf{v}}_i^t))$ is the image intensity at the projected image space position $\mathbf{\Pi}(\tilde{\mathbf{v}}_i^t)$ of 3D vertex $\tilde{\mathbf{v}}_i^t$ at time t, and $\nabla g_{t,i}$ is the corresponding spatial image gradient. Vertices at object boundaries and occlusions, which pose problems in 2D optical flow, are detected by projecting the template into both the camera and projector space, and checking each vertex for visibility. We set the per vertex weight to $w_i^{opt} = 1$ if visible and $w_i^{opt} = 0$ otherwise.

The optical flow constraint is equivalent to the point-to-texture constraint. The gradient defines the 2D texture normal, $\mathbf{n}_i^{tex} \nabla g_{t,i}$, and the texture correspondence is fixed to the optimal step along the gradient from the current projected image space position:

$$\mathbf{c}_i^{tex} = \mathbf{\Pi}(\hat{\mathbf{v}}_i^t) - \frac{\nabla g_{t,i}}{\|\nabla g_{t,i}\|_2^2}(g_{t+1,i} - g_{t,i}). \qquad (8.16)$$

Optical flow is applied in a hierarchical fashion using a 3-level Gaussian pyramid, where low resolution video frames are processed first to allow for larger deformations. In each optimization step, we re-project all visible vertices to the image plane and recalculate the spatial image gradient using a standard Sobel filter and the temporal derivative of the image intensity using forward differences.

(a) Geometry of manipulated object (left). Textured version (right).

(b) Backside of object (left). Scanned backside of object (right).

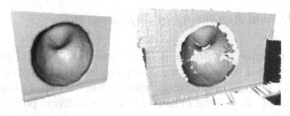

(c) Deformed object (left). Scanned deformed object (right).

Fig. 8.3 Example of deformation tracking on a hollow elastic half sphere mounted on a support plate

To improve convergence in the deformation recording, we schedule five optimization steps for each input scan by recalculating closest points and using coarse-to-fine video frame resolutions for optical flow. After rigid alignment, we perform three steps of optimization with increasing resolution in the Gaussian pyramid for estimating image gradients and two optimizations at the highest resolution. Each optimization step minimizes the total energy $E_{tot} = \alpha_{fit} E_{fit} + \alpha_{opt} E_{opt} + \alpha_{bend} E_{bend}$ with constant energy weights $\alpha_{fit} = 1$, $\alpha_{opt} = 100$, and $\alpha_{bend} = 100$. For untextured objects we simply omit the optical flow term. Note that the first frame is assumed to be in the rest state, and the initial alignment can therefore be done with any rigid registration method [46].

Figure 8.3(a) shows the undeformed template mesh of an exemplary object that is manipulated by the user. Figure 8.3(b) displays the backside of the object and one input scan of the rest pose. Figure 8.3(c) presents the corresponding object when the user pushes against the halfsphere: the geometry bends inside. The left part of

Fig. 8.4 Undeformed object (*left*), displacement field (*middle*), resulting deformation (*right*)

Fig. 8.3(c) shows the corresponding deformed template mesh. As can be seen, the deformations of the surface are tracked.

The proposed deformation tracking method is suitable for arbitrary smooth deformations. It is, however, less suited to capture fully articulated objects. Algorithms for the registration of articulated objects have been proposed by Chang and Zwicker [47], which could be included into our proposed registration framework. The deformation vectors $\mathbf{d}_{i,t}$ of each time frame define the deformations of the object.

8.4.2 Data-Driven Rendering of Deformations

During the manipulation of deformable objects the object shape changes according to the user interactions. To address these shape changes in the rendering, we apply an interpolation technique similar to the one used in data-driven haptic rendering. The approach relies on scanned object surfaces for different positions of the tool indenting the surface.

In the following two different approaches to interpolate the recorded deformations are described. The first one is suitable for small deformations in the proximity of the manipulated region, while the second method targets complex large-scale deformations of the whole object. The key task is to interpolate the observed deformation fields with respect to the contact point and tool position. In the following we describe the mathematical representation of deformation fields and then present the two interpolation schemes for arbitrary tool positions and contact points.

Representing a Deformation

Any deformation of an object can be represented by a vector field that describes the displacement of each object point. In Fig. 8.4 an undeformed cube is shown. If the depicted displacement field is applied, the deformed object is obtained. Similar to the interpolation technique used for haptic feedback, radial basis functions are employed to represent the deformation field. The displacement $\mathbf{d}$ of an object point $\mathbf{x}$ can be described as

$$\mathbf{d}(\mathbf{x}) = \sum_{i=1}^{N} \mathbf{w}_i \cdot \phi(\|\mathbf{x} - \mathbf{c}_i\|), \tag{8.17}$$

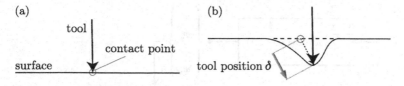

Fig. 8.5 Local deformation concept

where ϕ denotes the radial basis function, c_i are the centers, and w_i the weight vectors of the RBFs. We refer to this as the displacement interpolator or **d**-interpolator. The RBF centers c_i are selected in an initial step using a Gaussian random distribution around the contact point of the tool and the object surface. Moreover, the basis function ϕ is also selected in advance, thus the only remaining unknowns are the weight vectors w_i. The latter are obtained by solving a least-squares problem.

$$
\begin{pmatrix}
\phi(\|\mathbf{x}_1 - \mathbf{c}_1\|) & \cdots & \phi(\|\mathbf{x}_1 - \mathbf{c}_N\|) \\
\phi(\|\mathbf{x}_2 - \mathbf{c}_1\|) & \cdots & \phi(\|\mathbf{x}_2 - \mathbf{c}_N\|) \\
\vdots & \ddots & \vdots \\
\phi(\|\mathbf{x}_K - \mathbf{c}_1\|) & \cdots & \phi(\|\mathbf{x}_K - \mathbf{c}_N\|)
\end{pmatrix}
\cdot
\begin{pmatrix}
\mathbf{w}_1 \\
\mathbf{w}_2 \\
\vdots \\
\mathbf{w}_N
\end{pmatrix}
=
\begin{pmatrix}
\mathbf{d}_1 \\
\mathbf{d}_2 \\
\vdots \\
\mathbf{d}_K
\end{pmatrix},
\qquad (8.18)
$$

where $\mathbf{d}_i$ is the observed displacement of the surface point $\mathbf{x}_i$. Once the N weight vectors $\mathbf{w}_i$ are determined, the displacement field is completely specified.

During the recording phase the user performs different deformations for different indentations with the interaction tool at different contact points. Thus, multiple weight vectors $(\mathbf{w}_1, \ldots, \mathbf{w}_N)$ are required, each representing one specific deformation of the object. In the following, two approaches are presented that interpolate these deformations with respect to tool position and contact point.

Interpolation of Local Deformations

In the first approach it is assumed that the object does not undergo any large-scale deformations. Moreover, the local deformation should be independent of the location of the contact point. Hence, this approach only produces correct results for simple homogeneous planar or spherical objects. However, in practice it could also be applied to appropriately shaped local geometries of objects. Poor accuracy has to be expected at sharp object boundaries where the surface normal changes rapidly.

An example deformation is shown in Fig. 8.5. On the left, a tool is depicted upon contact with the object surface. On the right, it is pushed down and displaced laterally. The position of the tool tip with respect to the local coordinate frame at the contact point is denoted δ. For each position δ_i during a recording we calculate the set of weight vectors $\mathbf{W} = (\mathbf{w}_1, \ldots, \mathbf{w}_N)$ that specifies the corresponding displacement field. The task then is to find a weight vector set that corresponds to an arbitrary tool position δ.

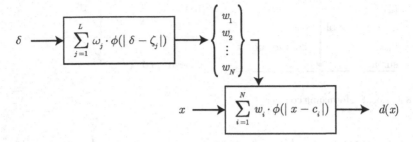

Fig. 8.6 Schematic representation of the interpolation of local deformations

To this end, we interpolate the weight vectors of the observed displacement fields with respect to the tool position. For the interpolation also RBFs are employed (denoted as δ-interpolator). Supposing deformations were recorded for some tool positions $\delta_1, \ldots, \delta_M$ yielding corresponding weight vectors $\mathbf{W}_1, \ldots, \mathbf{W}_M$ describing the individual deformations, then the interpolator is given by

$$\mathcal{W}(\delta) = \sum_{j=1}^{L} \omega_j \cdot \phi(\|\delta - \zeta_j\|), \tag{8.19}$$

where ω_j are the RBF weights and ζ_j the RBF centers of the δ-interpolator. Again, an appropriate set of centers needs to be chosen. Currently, we use all observed tool positions as centers. However, for larger data sets the number of centers has to be reduced, for instance by selecting only a subset of the observed tool positions. The weights ω_j are determined via the system

$$\begin{pmatrix} \phi(\|\delta_1 - \zeta_1\|) & \cdots & \phi(\|\delta_1 - \zeta_L\|) \\ \phi(\|\delta_2 - \zeta_1\|) & \cdots & \phi(\|\delta_2 - \zeta_L\|) \\ \vdots & \ddots & \vdots \\ \phi(\|\delta_M - \zeta_1\|) & \cdots & \phi(\|\delta_M - \zeta_L\|) \end{pmatrix} \cdot \begin{pmatrix} \omega_1 \\ \omega_2 \\ \vdots \\ \omega_L \end{pmatrix} = \begin{pmatrix} \mathcal{W}_1 \\ \mathcal{W}_2 \\ \vdots \\ \mathcal{W}_M \end{pmatrix}. \tag{8.20}$$

Once the weights ω_j are determined the displacement field for any tool position δ can be interpolated by evaluating the interpolator at the corresponding point. The overall process is illustrated in Fig. 8.6.

In the first step, the weight vector set $\mathbf{W}$ is determined. It is used in the second interpolation step to determine the displacement at any object point $\mathbf{x}$. As mentioned we assume the same local deformation behavior at arbitrary surface point. This requires the displacements as well as the values of $\mathbf{x}$ and $\mathbf{c}_i$ to be expressed independent of the contact point. To this end, we use a local coordinate frame at the contact point. The latter represents the origin, and the surface normal gives one of the coordinate axes. The tangential directions are generated from texture coordinates.

The points $\mathbf{x}$ for which we evaluate the displacement interpolator lie on the object surface. This would allow us to perform the interpolation in texture coordinates.

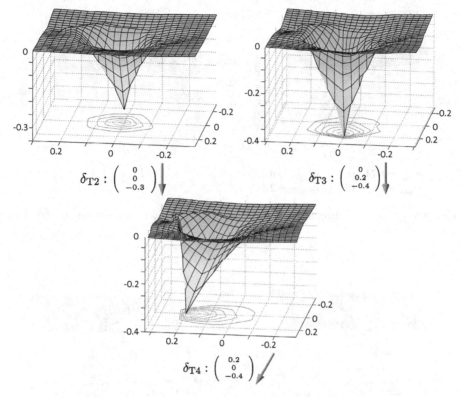

Fig. 8.7 Set of observations used to train the local deformation interpolator

However, for arbitrary surfaces it is difficult to map 3D to 2D coordinates while maintaining geometric distances. Points that lie very close in texture coordinates could be far away from each other in 3D space and vice versa. As RBFs rely on such distances to determine the interpolated value, we apply the **d**-interpolator in 3D space.

Considering polyharmonic splines as basis functions, the triharmonic spline $\phi = r^3$ is used to fit three-dimensional data [48]. However, the thin-plate spline $\phi = r^2 \log(r)$ showed slightly better results than the former for our data sets. This might be due to the fact that all tested points and centers were located on a plane with zero z-coordinate, yielding the same values $\mathbf{r} = \|\mathbf{x}_i - \mathbf{c}_i\|$ as for 2D coordinates.

As a first evaluation of the interpolation accuracy, we tested the algorithm on synthetic deformation data from a linear FEM model. Figure 8.7 shows three example deformations of a surface. Figure 8.8 illustrates the interpolated deformation for an unobserved tool position. The errors of the displacement vectors remain reasonable.

We performed the same test using a nonlinear material in the FEM simulation. For the training data the same tool positions as for the linear model were used. Figure 8.9 shows on the top an example of an interpolated deformation using a tool position from the training data. However, for unobserved positions the error

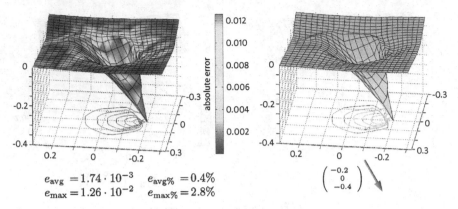

$$e_{\text{avg}} = 1.74 \cdot 10^{-3} \quad e_{\text{avg}\%} = 0.4\%$$
$$e_{\text{max}} = 1.26 \cdot 10^{-2} \quad e_{\text{max}\%} = 2.8\%$$

Fig. 8.8 Interpolated deformation error (*left*) and reference deformation (*right*) for $\delta = (-0.20 - 0.4)$

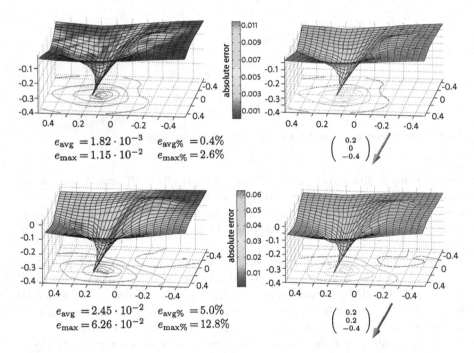

$$e_{\text{avg}} = 1.82 \cdot 10^{-3} \quad e_{\text{avg}\%} = 0.4\%$$
$$e_{\text{max}} = 1.15 \cdot 10^{-2} \quad e_{\text{max}\%} = 2.6\%$$

$$e_{\text{avg}} = 2.45 \cdot 10^{-2} \quad e_{\text{avg}\%} = 5.0\%$$
$$e_{\text{max}} = 6.26 \cdot 10^{-2} \quad e_{\text{max}\%} = 12.8\%$$

Fig. 8.9 Interpolated deformation error (*left*) and reference deformation (*right*) using a tool position from the training set as well as a new position

increases, as seen on the bottom example. The accuracy can be increased if more deformations are observed and used for the training of the data-driven model.

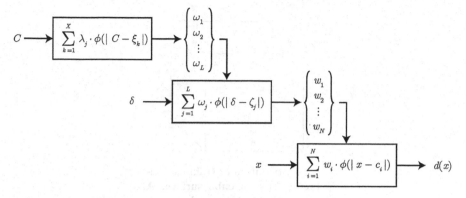

Fig. 8.10 Global deformation interpolation procedure

Interpolation of Global Deformations

In this approach the deformation field is not only dependent on the tool position δ, but also on the contact point $\mathbf{C}$. The general idea is to add a further interpolation step to the local deformation method. While previously we had one set of weights $\boldsymbol{\Omega} = (\omega_1, \ldots, \omega_L)$ for the tool position interpolator we now have an individual set of weights for each observed contact point. Using RBF interpolation we can again obtain the set of weights for any contact point $\mathbf{C}$. Each weight ω_i belongs to a specific center ζ_i. Hence, the same tool position centers ζ_i need to be used for each observed contact point in order to interpolate the corresponding weights ω_i. Therefore, we have to choose a set of tool positions as centers that we use for all contact points. We do not have a simple strategy, but experimental results suggest that the resulting accuracy is not sensitive to this choice, especially when dealing with linear materials. As long as the centers cover the range of performed indentations the interpolator is capable of reproducing the corresponding deformations. To acquire the weights $\boldsymbol{\Omega}$ for any contact point, we evaluate

$$\boldsymbol{\Omega}(\mathbf{C}) = \sum_{k=1}^{X} \lambda_j \cdot \phi(\|\mathbf{C} - \boldsymbol{\xi}_j\|), \tag{8.21}$$

where X is the number of weights λ_j and centers ξ_j, which we choose to be equal to the number of the observed contact points. We call this interpolator the contact point interpolator or $\mathbf{C}$-interpolator. The whole procedure to obtain the displacement for any contact point $\mathbf{C}$, any tool positions δ at any surface point $\mathbf{x}$ is depicted in Fig. 8.10.

Again, we tested the approach with synthetic data obtained in an FEM simulation of a linear elastic material. As a sample object we used a cube that was deformed at 45 different contact points. At each of these points different training load cases were applied (see Fig. 8.11). Figure 8.12 illustrates some of the training deformations for the tool position δ_2.

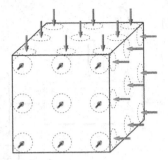

(a) Distribution of training contact points on cube surface. A regular 3×3 grid is used on each face except the bottom one.

$$\delta_1 = \begin{pmatrix} 0 \\ 0 \\ 0 \end{pmatrix}, \quad \delta_2 = \begin{pmatrix} 0 \\ 0 \\ -0.3 \end{pmatrix}\downarrow, \quad \delta_3 = \begin{pmatrix} -0.1 \\ 0 \\ -0.2 \end{pmatrix}\searrow, \quad \delta_4 = \begin{pmatrix} 0 \\ -0.1 \\ -0.2 \end{pmatrix}\downarrow$$

(b) Tool positions recorded at each contact point.

Fig. 8.11 Training load cases for interpolation of global deformation

Figure 8.13 shows an example deformation that was not used in the training of the interpolator. The top left figure illustrates the deformed cube as it was computed by the FEM simulation. In the top right figure the result from the interpolator is depicted. As can be seen in the plot, the error for the linear material is reasonable and the interpolated deformation is similar to the deformation obtained from the FEM.

8.5 Integrated System

The individual components introduced in the previous sections have been integrated into a first prototype for data-driven visuo-haptic acquisition and rendering of deformable bodies. Figure 8.14 gives an overview of the integrated system. The scheme illustrates the individual steps and signals used to recover and model the visual and haptic object behavior. Integrating the haptic recording setup with the visual capturing system required dealing with a number of challenges. These will be addressed in the following.

8.5.1 Synchronization and Calibration

The first point to consider is the temporal synchronization of the acquired data streams. Visual and haptic recording are performed by different processes, possibly

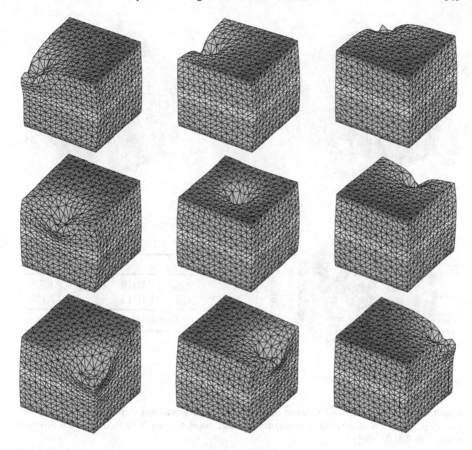

Fig. 8.12 Subset of training cases for training of global deformations

running on different machines. Hence, a temporal synchronization between the two capturing devices had to be ensured. One option to achieve this is using a hardware trigger signal.

In our prototype setup, the visual system triggered a signal whenever the cameras acquired an image, i.e. every 33 ms. The signal was received by the haptic recording device which sampled the trigger at 1 kHz. Subsequently, tool positions at the time of a signal were extracted for use in the data-driven deformation model.

Besides the temporal synchronization the two systems also need a common coordinate frame in which both the 3D geometry as well as the tool interaction can be described. An easy solution is the use of a physical calibration geometry that is fixed in the workspace of the recording devices.

In our setup we used a checkerboard for this purpose. The board was first visually recognized in images acquired with the visual system. This allowed an automatic computation of the corresponding coordinate frame. Next, the haptic system was calibrated by moving the device tool tip to several points on the checkerboard.

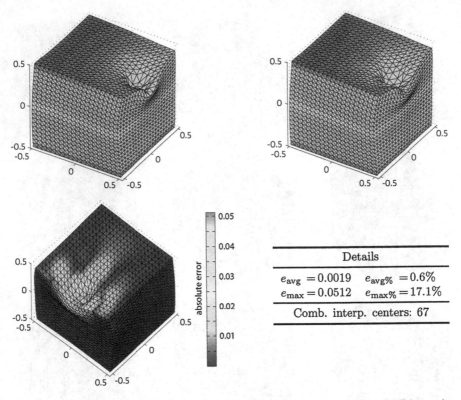

Fig. 8.13 Test interpolation of global deformation. FEM simulation (*top left*), RBF interpolation (*top right*), resulting error (*bottom*) for Contact point $C = (0, 0.39, 0.5)$ and tool position **delta** $= (0.2, 0.0, -0.4)$

For these locations the haptic tool coordinates as well as the corresponding 3D coordinates given by the visual system were stored. Using these two 3D datasets the transformation between the haptic and the visual coordinate frame was determined.

The determined transformation allows to align the geometric object mesh with the haptic recordings. In the current prototype, the mesh was used to detect tool collisions with the object. Based on the surface normal and texture coordinates, a contact coordinate frame was established. Tool trajectory as well as acquired force data were transformed into this contact frame. In addition to this, the cross-calibration was used to align the tool trajectory with the measured deformation data. As depicted in Fig. 8.14 these two data streams were employed to compute the data-driven deformation model. Since the visual system captured data at 30 Hz while the haptic system employed 1 kHz sampling, the deformation was not observed for every sampled tool position. Hence, only tool positions were used for which the visual system issued a trigger signal.

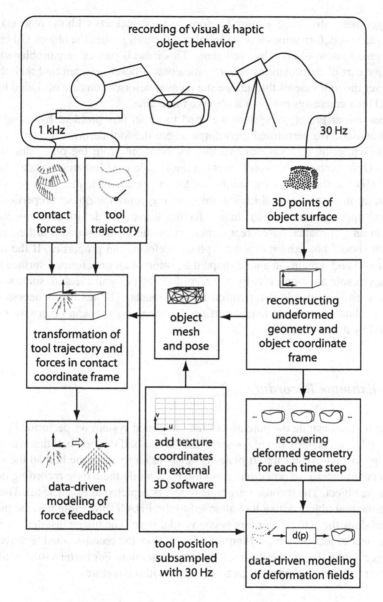

Fig. 8.14 Overview of the integrated visuo-haptic recording system and the signal flow

8.5.2 System Limitations

A major difficulty encountered during the prototype integration of the visual and haptic system were occlusions caused by the probing tool. In order to reconstruct the object geometry as well as deformations, the contact area between the object and the tool must be fully observable by the camera system. To this end, the undeformed object geometry should be acquired before the user interacts with the object. However, regarding deformations it is essential that the user probes the object. Otherwise no deformations would be observed at all. This issue is particularly problematic as the major part of deformations occur at the contact point between tool and object. Whenever the tool indents the surface the created deformations are occluded by the tool and the camera system cannot observe these data.

In our initial prototype system we tried to avoid this problem by using thin-shelled objects. We performed recordings where the tool deforms the object from one side while the cameras observe the deformations from the other side of the surface. Unfortunately, this solution is not applicable to volumetric objects. Nevertheless, this test showed the capabilities of the integrated system.

In addition to this, several limitations exist regarding the object properties. The modeling approach used for the haptic feedback assumes deformable objects and relies on an impedance-based representation of the virtual scene. Further, object behavior should not exhibit extensive plastic deformation properties. If the object is deformed and remains in a new shape the notion of an undeformed surface mesh is not applicable anymore. Finally, the recorded object should feature some surface texture to ensure reliable deformation measurements. The texture is necessary to obtain a robust estimate of lateral surface deformations that can otherwise not be observed by the camera system.

8.5.3 *Example Recording*

In order to demonstrate the potential of the integrated system we performed several test recordings. As one example object we used one half of a toy ball that was glued on a support plate. The support plate was cut out underneath the ball so the visual system could record deformations from the back while the haptic recording device probed the object. The recorded example object is depicted in Fig. 8.15(a). The user indents the real object with a tool attached to the PHANToM. Note that the picture does not show the actual recording session and device. The image just demonstrates how the real object deformed. Figure 8.15(b) shows the reconstructed textured virtual object while the user performs a similar indentation. For better visualization of the deformation, Fig. 8.15(c) depicts the scene without texture.

8.6 Conclusions and Discussion

In this chapter we presented the individual components of our data-driven visuo-haptic acquisition and rendering system. The visual as well as haptic acquisition and display frameworks have been introduced. We covered the required extensions to also process deformable objects with the presented system. We outlined our prototype framework and addressed the various steps required for optimal integration.

(a) Indentation of the real object with a normal stylus.

(b) Rendering of the recorded virtual object for a similar indentation.

(c) Rendering of the virtual representation without texture.

Fig. 8.15 Example object that was recorded with the integrated visuo-haptic system

We envision the use of this technology for instance in cases where objects need to be presented to remotely located persons. An example for this is product development involving geographically distributed teams. In this context, preliminary prototypes might have to be presented to collaborators thousands of miles away. As the physical transportation of the object is too slow, expensive, and potentially logistically complex, the presentation process usually relies on electronic communication.

The developed record/replay technology offers a solution to allow full multi-modal exploration of the object. During the recording process all necessary look-and-feel information about the object of interest are fully encoded in an electronic representation, and can be instantaneously transmitted to anywhere in the world. The replay/create environment then enables the free visuo-haptic exploration of a transmitted, high-quality virtual copy of the object. The presented work represents a first step towards this future goal.

Acknowledgements This work was partly supported by the ImmerSence project within the 6th Framework Programme of the European Union, FET—Presence Initiative, contract number IST-2006-027141, see also www.immersence.info.

References

1. Hoever, R., Harders, M., Szekely, G.: Data-driven haptic rendering of visco-elastic effects. In: IEEE Haptic Symposium, pp. 201–208 (2008)
2. Hoever, R., Kosa, G., Szekely, G., Harders, M.: Data-driven haptic rendering-from viscous fluids to visco-elastic solids. IEEE Trans. Haptics **2**, 15–27 (2009)
3. Hoever, R., Di Luca, M., Szekely, G., Harders, M.: Computationally efficient techniques for data-driven haptic rendering. In: World Haptics, pp. 39–44 (2009)
4. Weise, T., Leibe, B., Van Gool, L.: Fast 3D scanning with automatic motion compensation. In: CVPR07, pp. 1–8 (2007)
5. Weise, T., Wismer, T., Leibe, B., Van Gool, L.: In-hand scanning with online loop closure. In: IEEE International Workshop on 3-D Digital Imaging and Modeling (2009)
6. Scharstein, D., Szeliski, R.: A taxonomy and evaluation of dense two-frame stereo correspondence algorithms. Int. J. Comput. Vis. **47**(1–3), 7–42 (2002)
7. Scharstein, D., Szeliski, R.: http://vision.middlebury.edu/stereo/

8. Klaus, A., Sormann, M., Karner, K.F.: Segment-based stereo matching using belief propagation and a self-adapting dissimilarity measure. In: ICPR, pp. 15–18 (2006)
9. Blais, F.: Review of 20 years of range sensor development. In: Videometric VII, Proceedings of SPIE Electronic Imaging, vol. 5013, pp. 62–76 (2003)
10. Lange, R., Seitz, P., Biber, A., Schwarte, R.: Time-of-flight range imaging with a custom solid state image sensor. Laser Metrol. Inspect. **3823**(1), 180–191 (1999)
11. Batlle, J., Mouaddib, E., Salvi, J.: Recent progress in coded structured light as a technique to solve the correspondence problem: A survey. Pattern Recognit. **31**(7), 963–982 (1998)
12. Salvi, J., Pagès, J., Batlle, J.: Pattern codification strategies in structured light systems. Pattern Recognit. **37**(4), 827–849 (2004)
13. Scharstein, D., Szeliski, R.: High-accuracy stereo depth maps using structured light. Comput. Vis. Pattern Recognit. **01**, 195–202 (2003)
14. Wust, C., Capson, D.W.: Surface profile measurement using color fringe projection. Mach. Vis. Appl. **V4**(3), 193–203 (1991)
15. Zhang, L., Curless, B., Seitz, S.M.: Spacetime stereo: Shape recovery for dynamic scenes. In: IEEE Computer Society Conference on Computer Vision and Pattern Recognition, pp. 367–374 (2003)
16. Davis, J., Nehab, D., Ramamoorthi, R., Rusinkiewicz, S.: Spacetime stereo: A unifying framework for depth from triangulation. IEEE Trans. Pattern Anal. Mach. Intell. **27**(2), 296–302 (2005)
17. Koninckx, T.P., Griesser, A., Van Gool, L.J.: Real-time range scanning of deformable surfaces by adaptively coded structured light. In: 3DIM, pp. 293–301 (2003)
18. Zhang, L., Curless, B., Seitz, S.M.: Rapid shape acquisition using color structured light and multi-pass dynamic programming. In: The 1st IEEE International Symposium on 3D Data Processing, Visualization, and Transmission, pp. 24–36 (2002)
19. MacLean, K.: The 'haptic camera': A technique for characterizing and playing back haptic properties of real environments. In: Proc. of ASME Dynamic Systems and Control Devision, vol. 58, pp. 459–467 (1996)
20. Greenish, S., Hayward, V., Steffen, T., Chial, V., Okamura, A.: Measurement, analysis, and display of haptic signals during surgical cutting. Presence **11**, 626–651 (2002)
21. Okamura, A., Webster, R., Nolin, J., Johnson, K., Jafry, H.: The haptic scissors: Cutting in virtual environments. In: Proc. of the ICRA, vol. 1, pp. 828–833 (2003)
22. Edmunds, T., Pai, D.K.: Perceptual rendering for learning haptic skills. In: IEEE Haptic Symposium, pp. 225–230 (2008)
23. Colton, M., Hollerbach, J.: Reality-based haptic force models of buttons and switches. In: Proc. of the ICRA, pp. 497–502 (2007)
24. Colton, M., Hollerbach, J.: Haptic models of an automotive turn-signal switch: Identification and playback results. In: Proc. of the Second Joint EuroHaptics Conference and Symposium on Haptic Interfaces for Virtual Environment and Teleoperator Systems, pp. 243–248 (2007)
25. Kry, P., Pai, D.: Interaction capture and synthesis. ACM Trans. Graph. **25**, 872–880 (2006)
26. Andrews, S., Lang, J.: Interactive scanning of haptic textures and surface compliance. In: Proc. of the International Conference on 3-D Digital Imaging and Modeling, pp. 99–106 (2007)
27. Pai, D.K., Rizun, P.: The WHaT: a wireless haptic texture sensor. In: IEEE Haptic Symposium, pp. 3–9 (2003)
28. Richard, C., Cutkosky, M., MacLean, K.: Friction identification for haptic display. In: Proc. of the ASME Dynamic Systems and Control Division, vol. 67, pp. 327–334 (1999)
29. Kuchenbecker, K., Fiene, J., Niemeyer, G.: Event-based haptics and acceleration matching: Portraying and assessing the realism of contact. In: Proceedings of the First Joint Eurohaptics Conference and Symposium on Haptic Interfaces for Virtual Environment and Teleoperator Systems, pp. 381–387 (2005)
30. Kuchenbecker, K., Fiene, J., Niemeyer, G.: Improving contact realism through event-based haptic feedback. In: IEEE Transactions on Visualization and Computer Graphics, vol. 12, pp. 219–230 (2006)
31. Pai, D., Lang, J., Lloyd, J., Woodham, R.: ACME, a telerobotic active measurement facility. In: Experimental Robotics VI, vol. 250, pp. 391–400 (2000)

32. Pai, D., van den Doel, K., James, D., Lang, J., Lloyd, J., Richmond, J., Yau, S.: Scanning physical interaction behavior of 3D objects. In: ACM SIGGRAPH 2001 Conference Proc., pp. 87–96 (2001)
33. Sedef, M., Samur, E., Basdogan, C.: Visual and haptic simulation of linear viscoelastic tissue behavior based on experimental data. In: Haptic Symposium, pp. 201–208 (2006)
34. Samur, E., Sedef, M., Basdogan, C., Avtan, L., Duzgun, O.: A robotic indenter for minimally invasive characterization of soft tissues. In: Proc. of the Computer Assisted Radiology and Surgery, vol. 1281, pp. 713–718 (2005)
35. Mahvash, M., Hayward, V.: Haptic simulation of a tool in contact with a nonlinear deformable body. In: Surgical Simulation and Soft Tissue Deformation, vol. 2673, pp. 311–320 (2003)
36. Mahvash, M., Hayward, V.: High fidelity haptic synthesis of contact with deformable bodies. IEEE Comput. Graph. Appl. **24**, 48–55 (2004)
37. Allen, B., Curless, B., Popović, Z.: The space of human body shapes: reconstruction and parameterization from range scans. In: ACM SIGGRAPH 2003 Papers. SIGGRAPH '03, pp. 587–594 (2003)
38. Amberg, B.: Optimal step nonrigid ICP algorithms for surface registration. In: CVPR '07 (2007)
39. Li, H., Sumner, R.W., Pauly, M.: Global correspondence optimization for non-rigid registration of depth scans. Comput. Graph. Forum **27**(5) (2008)
40. Zhang, L., Snavely, N., Curless, B., Seitz, S.M.: Spacetime faces: High-resolution capture for modeling and animation. In: ACM Annual Conference on Computer Graphics, pp. 548–558 (2004)
41. Botsch, M., Sorkine, O.: On linear variational surface deformation methods. IEEE Trans. Vis. Comput. Graph. **14**, 213–230 (2008)
42. Schenk, O., Gärtner, K.: Solving unsymmetric sparse systems of linear equations with PARDISO. Future Gener. Comput. Syst. **20**, 475–487 (2004)
43. Mitra, N.J., Gelfand, N., Pottmann, H., Guibas, L.: Registration of point cloud data from a geometric optimization perspective. In: Proceedings of the 2004 Eurographics/ACM SIGGRAPH Symposium on Geometry Processing, pp. 22–31 (2004)
44. Horn, B.K.P., Schunck, B.G.: Determining optical flow. Artif. Intell. **17**, 185–203 (1981)
45. Decarlo, D., Metaxas, D.: Optical flow constraints on deformable models with applications to face tracking. Int. J. Comput. Vis. **38**, 99–127 (2000)
46. Weise, T., Leibe, B., Van Gool, L.: Accurate and robust registration for in-hand modeling. In: Computer Vision and Pattern Recognition, pp. 1–8 (2008)
47. Chang, W., Zwicker, M.: Automatic registration for articulated shapes. In: Proceedings of the Symposium on Geometry Processing, pp. 1459–1468 (2008)
48. Iske, A., Arnold, V.I.: Multiresolution Methods in Scattered Data Modelling. Springer, Berlin (2004)

Chapter 9
Haptic Rendering Methods for Multiple Contact Points

Manuel Ferre, Pablo Cerrada, Jorge Barrio, and Raul Wirz

Abstract Vast majority of haptic applications are focused on single contact point interaction between the user and the virtual scenario. One contact point is suitable for haptic applications such as palpation or object exploration. However, two or more contact points are required for more complex tasks such as grasping or object manipulation. Traditionally, when single-point haptic devices are applied in a complex task, a key or a switch is used for grasping objects. Using multiple contact points for this kind of complex manipulation tasks significantly increases the realism of haptic interactions. Moreover, virtual scenarios with multiple contact points also allow developing multi-user cooperative virtual manipulation since several users simultaneously interact over the same scenario and perceive the actions being performed by others. It represents a step forward in the current haptic applications that are usually based on one single user.

Recreating these scenarios in a stable and realistic way is a very challenging goal due to the complexity of the computational models that requires integrating all interactions of multiple haptic devices and calculations of the corresponding actions over the virtual object scenarios. It also requires solving mathematical equations in real time that properly represent the behavior of virtual objects. Delays in these calculations can lead to instabilities that may produce vibrations in the haptic devices and reduce the realism of the simulations. In order to offset these problems, different kinds of solutions have been considered, such as (i) models based on simplified calculations of forces and torques that are involved in object interactions, (ii) models based on virtual coupling between objects in order to ensure stability of the simulation.

M. Ferre (✉) · P. Cerrada · J. Barrio · R. Wirz
Centre for Automation and Robotics UPM-CSIC, Universidad Politecnica de Madrid,
C/. Jose Gutierrez Abascal, 2, 28006 Madrid, Spain
e-mail: m.ferre@upm.es

P. Cerrada
e-mail: pablo.cerrada@upm.es

J. Barrio
e-mail: jordi.barrio@upm.es

R. Wirz
e-mail: r.wirz@upm.es

A. Peer, C.D. Giachritsis (eds.), *Immersive Multimodal Interactive Presence*,
Springer Series on Touch and Haptic Systems,
DOI 10.1007/978-1-4471-2754-3_9, © Springer-Verlag London Limited 2012

Experiments shown in this work have been performed by using a multifinger haptic device called MasterFinger-2(MF-2). Is has been applied in different kinds of multiple contact point applications.

9.1 Introduction

Haptic devices are mechanical devices designed to exert forces to a human user in order to create the sensation of touch or the manipulation of real objects. Furthermore, they do not only reflect forces to a person, but also capture the position and movements of the hand and/or fingers. Combining these haptic devices with visual interfaces allows the operator's perception to be highly enhanced and thereby improve the performance of certain tasks [1].

Haptic interfaces can be used for different kind of applications. Firstly, a haptic device can act as a master of a robot manipulator [2, 3]. In such manner, the haptic device will reproduce the forces applied by the robot to the operator. Secondly, other types of application take place when the haptic device is used for interacting with virtual scenarios [4]. These virtual scenarios are freely designed and focus on specific tasks; therefore, new kinds of uses are continuously emerging, such as training [5], medical rehabilitation [6], prototype designing and many others. Finally, augmented reality applications combine real and virtual scenarios [7, 8]. In this case, a real scenario is enhanced by the generation of virtual interactions.

In this chapter, the main focus is on the haptic devices that work in virtual scenarios. In this case, instead of interacting with real scenarios and measuring forces, a mathematical model of the object to be touched is considered. A model that gathers its representative features related to its dynamics and to its haptic sensations. Combining both virtual scenario and haptic devices, different haptic rendering algorithms have been developed. These algorithms are responsible for calculating forces exerted to the operator during manipulation of the virtual scenarios, taking into account all dynamics and friction properties of the objects.

So far, most haptic devices that are commercially available have been firstly conceived and developed in order to allow single contact-point interaction between the user and the scenario [9]. There are plenty of interesting tasks that can be simulated by means of only one contact point, such as explorations, palpations, or many other interactions. Furthermore, these haptic devices together with their specific haptic rendering methods have reached very high levels of performance and realism as far as single contact point interactions are concerned.

However, there are many other applications where more than one contact point is required, and having haptic feedback available can greatly improve the performance. Some applications may require to simulate complex manipulations, such as grasping of delicate objects or other tasks that need simultaneous cooperation among users, etc.

The development of these virtual interactions involving multiple contact points is not a trivial problem in terms of hardware, nor of software. Most of the times, when trying to put together two or more conventional single contact-point devices, some problems arise, such as collisions among mechanical structures, consequently reducing usable workspace and making the whole integration nonfunctional. Similar problems arise when trying to extend the existing rendering algorithms to multi contact-point scenarios. Frequently, these algorithms are not capable of managing all interactions and thereby problems of stability arise.

In order to come up with solutions for these problems, several experiments have been developed by using the MasterFinger-2 haptic device. These experiments have been designed and carried out under the IMMERSENCE project by the UPM in collaboration with other partners. This haptic interface was used in different applications with multiple contact points, such as weight perception discrimination [10, 11], cooperated manipulation tasks by two users, and multi-modal interaction evaluation in grasping tasks [12, 13].

In the following section, some of the most powerful methods for multiple contact-point rendering are summarized. Then, main requirements for implementing those methods are reviewed and the corresponding software architecture resulting from those requirements is explained. Finally, some experiments performed by using the MasterFinger-2 device are shown.

9.2 Methods for Multiple Contact Points

Since haptic devices began to appear in the early 1990s, many different haptic rendering algorithms have been developed with very diverse purposes: trying to maximize stiffness of rigid objects, trying to deal with deformable bodies, etc. [14]. This chapter is centered to rendering methods, which are aimed at the recreation of multiple contact points between rigid objects and the operator.

Despite the wide range of applications, it can be easily observed in the literature that nearly all haptic rendering methods share the same basic structure, based on the three blocks illustrated in Fig. 9.1: "Collision detector", "Force calculator" and "Haptic device controller" [15].

Collision-detection algorithms are responsible for checking the contacts occurring between the virtual objects in the scenario and the virtual representation of the end-user, also known as "avatar" [15] or "virtual tool" [16].

Force calculation algorithms work when a collision has been detected, returning interaction force between the virtual representation of the end-user and the virtual objects. In the haptic device control block, low-level control loops are running. They control the actuators in the device so that the force calculated in the previous block is properly reflected to the user and updates the user's movements.

The factors that really make the difference among existing methods for haptic rendering are concentrated on the second block. They are not only responsible for calculating forces but also for the stability of the whole haptic simulation that relies on it.

To our concern, several approaches for computing the interaction forces between the avatar and the virtual objects can be considered. Most of them are derived from

Fig. 9.1 Basic structure of a
haptic rendering system

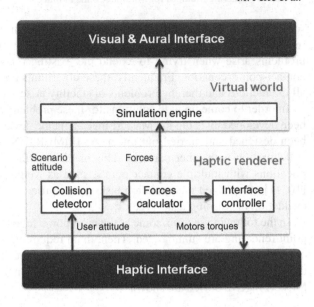

previous algorithms that have been originally developed for single-contact-point interactions and extended afterwards.

The first type of methods is based on force fields associated to the virtual objects [17]. These methods come from the former penalty methods based on the
spring approach [18, 19], which estimates the interaction force proportionally to
the registered penetration among objects. In a more general context, the force field
methods define forces according to geometrical and material properties associated
to each object. These methods often work well when dealing with static objects or
when only one contact point is involved in the interaction. However, these methods
have two main drawbacks when they are used for multiple contact-point manipulations of rigid bodies:

- These algorithms do not take into account past positions of the end-user, and
  thereby cause a problem when dealing with thin objects such as glass. Such problems arise since the virtual tool crosses thin surfaces when it moves too fast to be
  properly computed by the collision detector, or if the user simply pushes harder
  than the maximum strength of the haptic device.
- Force direction is also complex to determine during manipulation. The force
  fields are usually defined perpendicularly pointing to the exterior of the surface,
  and consequently, the interaction forces obtain zero tangential component that
  make the manipulation tasks hardly realistic or even difficult to perform.

In order to extend this method to multi-contact point applications, different approaches based on virtual coupling between the virtual representation of the user
and the objects were proposed in [20] and [21]. In [22], different grabbing conditions were defined for each object. Furthermore, depending on the relative positions
among the fingers and the virtual object to be grasped, a specific condition was
applied. Once the grasping condition is satisfied, the virtual representations of the

fingers and the virtual objects were coupled. Then, virtual fingers and the object would move together until the grasping condition ends.

A second group of methods developed under a different approach are those derived from the 'god-object' or proxy paradigms. These algorithms were first introduced by Zilles et al. in [23] and by Ruspini et al. in [24]. Unlike force field based algorithms, this family of algorithms does not only take into account the present position of the virtual tool, but also its trajectory. In such manner, the problems mentioned in the previous paragraphs could be avoided.

These algorithms have been designed based on the same mechanism: the god-object or proxy is a conceptual point restricted to the surface of the objects, and its position at every single moment is the closest to the virtual representation of every finger. As long as the user stays away from the object, no forces will be exerted. Once the user penetrates the virtual object, the force will be calculated proportionally to the distance between the proxy and the position of the user's virtual representation. This algorithm solved the first drawback that had been previously mentioned. Considering the tangential component problem, a more physical approach combining the proxy with a friction-cone model was proposed in [25, 26] by Melder et al. modifying the Zilles' method by making the god-object static once the user has penetrated the object. A force can be calculated using the spring approach between the god-object and the representation of every finger. In such manner, the total force has a tangential component. Applying this method to all the contact points in the simulation, the resultant force and the torque to the virtual object can be calculated. Thereby, allowing complex manipulation tasks to be performed. Hence, the object can be moved and rotated as desired.

9.3 Software Architecture for Multiple Contact Points in Haptics Rendering

Software architecture design is an important decision since simulation of multiple contact points implies a great number of very demanding processes running in real time. Some processes are related to the haptic devices, others to graphical simulations, and others to the physical engine that implements the virtual scenario. Hence, the design of the software architecture should be oriented to satisfying the application requirements. It is well known that for smooth transitions in graphics, updates of 50 Hz is required. So visual information should be refreshed at least 50 Hz so as to get realistic graphics. Meanwhile, in terms of force feedback, refresh rates significantly lower than 1 kHz, would produce vibrations felt by the user. Furthermore, the simulation engine has to be able to manage scenarios where complex physical entities are involved, such as fluids or soft bodies. Other key points in designing the software architectures are related to communication (what information is transmitted), distributed or centralized concept (how information is transmitted), and integration of processes with different frequency rates.

Fig. 9.2 Different layers for the distributed software architecture implemented by using the MasterFinger-2

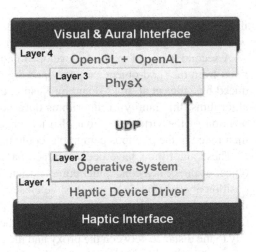

An ideal software architecture should be independent from the hardware as much as possible. Architecture software components should also be easy to install, upgrade and maintain on different platforms. As explained in Fig. 9.1, there are two main components in the software architecture that can be considered independent: drivers of the haptic device and the physical engine. The driver of a haptic device is a software component strongly related to the hardware and to the operative system, which makes the choice very little for this component. On the other hand, since the physical engine used for implementing the virtual scenario does not depend on the hardware of the haptic device, it allows a wide range of physical engines for selection. The selection is done according to several parameters, such as the possibility of accelerating the calculations through a GPU, the resolution of the engine, or the different physical entities that it supports.

Figure 9.2 describes the MasterFinger-2 software architecture. This software architecture shows a very similar shape to the generic one presented in Fig. 9.1, except in this case, the generic "haptic renderer" has been split into two different layers (Linux RTAI and VxWorks) in order to ease communications and so enable applications where more than one device is required. In general terms, in this architecture, each layer represents the chosen software package to implement the generic architecture. Below, each module is described:

- Layer 1: Haptic Device Driver: The driver is developed by the haptic device manufacturer and is the linkage between the haptic device and the operating system. For that reason, drivers are totally dependent on the hardware and are specific for each particular device and operating system. The MasterFinger-2 driver is implemented on a FPGA that runs on real time operating system called VxWorks.[1]
- Layer 2: Operative System: The choice of the operating system is done according to the previous layer. There are two operating systems that are commonly used for

---

[1]http://www.windriver.com/products/vxworks/.

commercial haptic devices: Linux and Windows. If a haptic device driver is supported by both operating systems, the selection of one or the other will depend on the final application. However, for haptic applications of real-time features are required for the OS. There are software packages that transform Windows or Linux in a real-time operating system. MasterFinger2 uses a Linux distribution (Ubuntu) with a patched kernel to obtain a real time operating system called RTAI.[2]

- Layer 3: Physics Motor: This layer is responsible for creating and updating the virtual scenario that the end user can interact with. This layer calculates interactions among virtual objects of the scenario, such as collisions, forces, velocities, positions or torques. In this case, different solutions can be also selected, such as Havok, PhysX or Bullet. In our case, PhysX[3] has been selected. The main reason is the possibility to use Nvidia GPU's since it allows much faster simulations.

- Layer 4: Graphics Interface: This layer represents the graphic library used for showing the virtual scenario to the user. It is relevant to mention that most of the haptic applications include a graphical visualization of the virtual scenario. In this particular case, OpenGL has been used since it is supported by PhysX, which is quite common in haptic applications.

The proposed software architecture is designed in a way that each layer is independent from the other and its communication is performed by using the UDP through a local LAN without congestion. However, if the architecture works over a non-local LAN or a LAN with congestion, a more sophisticated protocol can be used, such as the BTP protocol [27].

The main objective of these applications is to show how the operator is interacting with virtual objects while the operator is handling the haptic device. The simplest way to show that information is by using visual and haptic channels where the operator is stimulated with both kinds of information (visual and haptic) in order to properly perform the manipulation. Both stimuli channels have different features regarding frequencies and it is necessary to have proper synchronization of both stimuli, so a synchronization mechanism should then be required in order to maintain the coherence between both channels.

In principle, the force information is reproduced in the haptic interface (layer 1) and high frequency (about 1000 Hz) is required [28]. Layer 2 must support layer 1 data flow. On the other hand, the visual information is then displayed on the graphics interface (layer 4) and a frequency of 60 Hz is required (120 Hz if 3D stereoscopic). Layer 3 frequency is the same as layer 4, since a higher frequency will not be noticed by the end-user. So, a middle layer is necessary to join the two different layers. The middle layer should have a minimum of two buffers for compensating frequencies: One buffer for determining the real position of the haptic device (information from layer 1 to layer 3) and the force applied in the haptic device (information from layer 3 to layer 1).

---

[2]http://www.rtai.org/.

[3]http://developer.nvidia.com/object/physx.html.

Fig. 9.3 A user executing the
a uni-manual lift. Right hand
is used for grasping the box

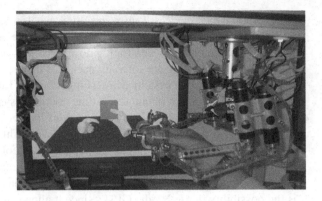

9.4 Bimanual and Cooperative Manipulation

MasterFinger-2 haptic device and the previously described architecture have been
applied in developing several cooperative manipulation scenarios. The first developed experiment is an approach to object grasping and weight perception. It was
designed for evaluating how realistic the virtual manipulation is. The second experiment presents a virtual scenario where two participants have to coordinate their
movements in order to lift a virtual prism together. Each participant did the same
experiment with and without force feedback.

9.4.1 Virtual Object Weight Discrimination

In this experiment, the haptic interface is used for lifting virtual weights [10, 11].
A virtual scenario was created in order to carry out this experiment. This scenario
sets a virtual box in the scenario that had to be lifted. The user is able to grasp the box
and lift it. The scenario includes a graphic interface where the finger movements and
the box are displayed on a screen, see Fig. 9.3. This virtual scenario always shows
the same kind of box, but with different weights. The scenario displays two boxes
with different weights, where the user has to decide which box is heavier according
to the virtual weight haptically displayed.

Forces delivered to the user are calculated by means of a penalty-based algorithm.
A force proportional to the penetration of the finger into the object is delivered to
the user when the collision is detected. When both fingers touch the virtual box, a
grasping phase is raised at the scenario server. During the grasping phase, the user
feels a force which is orthogonal to the surface of the box that corresponds to the
grasping force; if the grasping force is sufficient so as to avoid the box to slip, then
the box can be lifted. When the user begins to lift the box, an additional vertical
force that corresponds to the load of the virtual box is felt by the user. When the
box is lifted, its weight is shared by the thumb and the index finger (50% of load per
finger), since movement is restricted to the vertical axis.

According to the experiment results, MF-2 is five times less sensitive than real weight, there may be both user-specific and interface-specific reasons for these results. The users had no reliable cutaneous feedback about the weights and therefore relied only on proprioceptive information to estimate the virtual weight. The lack of cutaneous information could have resulted in a deterioration of weight sensitivity.

Nonetheless, the haptic device, MasterFinger-2, has managed to simulate weight sensation effectively and establish a clear presence for each of them. This kind of haptic interfaces with effective weight simulation offers professionals the opportunity to train in precision tasks with accurate weight perception. It also offers scientists the possibilities to investigate dynamic aspects of weight perception (e.g., sensitivity to weight changes during object displacement), which is very difficult or even impossible with real weights.

9.4.2 A Virtual Word Collaborative Task: Two Persons Lifting an Object

In order to get an approach to collaborative tasks between two end-users through haptic devices, an experiment based on the MasterFinger-2 was developed. In this experiment, two users cooperate in order to lift a virtual prism until it reaches a defined altitude from the floor and, then, the virtual prism has to be carefully placed back onto the floor. During the experiment, if both users did not coordinate their movements properly, the prism could be broken. In such case, the trial would be considered as failed. The main parameters of every trial have been recorded and the analysis of the data is presented in detail in the following paragraphs.

The experiment consists of two tests. The first test consists of a cooperative lifting of the prism without haptic information; the users perform the task by using only visual information provided by the scenario. The second is a similar test carried out by the same users that use not only visual but also haptic information. The main goal of this experiment is to measure the improvement obtained by using haptic information on a collaborative task. Both participants have to collaborate with each other in order to lift a deformable object. The experiment consists of two tests: One test that requires the participant to lift the virtual prism with visual and haptic information, and the other test requires the same participants to lift the virtual prism using visual information only. The object would break if participants applied more pressure (higher than 3.2 N). The object has to be lifted 50 times to an altitude higher than 20 cm and then placed back on the virtual ground.

Twenty-four students of the UPM volunteered in the experiment. They teamed up in twelve pairs and carried out the experiment. Each member of the team was asked to stay in different rooms in order to avoid any kind of communication between them. The only interaction allowed was through the collaborative scenario. Each member of the team used a haptic scenario made up of a MasterFinger2 and a monitor to visualize the representation of the collaborative task. The MasterFinger2 was mounted in a metallic structure upside-down, see Fig. 9.4.

Fig. 9.4 People performing a collaborative lifting of a prism. Each person is isolated from the other in a different room

The collaborative task manipulates a virtual prism which can break if it is not properly manipulated. Therefore, the effectiveness of a collaborative task is measured according to the number of successful tasks, broken objects and slipped objects. Figure 9.5 compares both types of collaborative tasks (a collaborative task with haptic data versus a collaborative task with visual information only). The results have shown that there is a bigger success rate, 92.5%, when the collaborative task is carried out with haptic data compared to a task carried out with visual information only, 83.3%. The visual collaborative task has more than twice the percentage of broken object 15.8%, compared to a haptic collaborative task which is 6.3%. The percentage of slipped objects is almost negligible for both kind of collaborative tasks. Although the percentage rates of a visual collaborative task are not very bad; a haptic collaborative task exhibits excellent percentage rates. It indicates that the haptic data is a key stimulus that allows performing a collaborative task more precisely and efficiently. Analyzing the results, the mean time spent for a haptic collaborative task is 9.989 seconds. The mean time spent for a visual collaborative task is 11.664 seconds. There is a significant mean time difference between both kinds of task, 1.675 seconds. It means that haptic stimuli are a key aspect for conducting the experiment with more precision and agility. This experiment compares haptic and visual collaborative tasks in order to evaluate the significance of haptic stimuli in collaborative tasks. According to the results from the experiment, it clearly shows that haptic stimuli have improved virtual collaborative tasks. It is confirmed that people prefer similar stimuli to the one perceived on real interactions, as the learning process gets accelerated, and the success rate in the required tasks is highly improved.

9.5 Conclusions

A new kind of virtual manipulation tasks can be carried out by using multiple contact point haptic devices. In this chapter, some examples such as grasping and cooperative manipulation have been shown. The cooperative manipulation scenario has

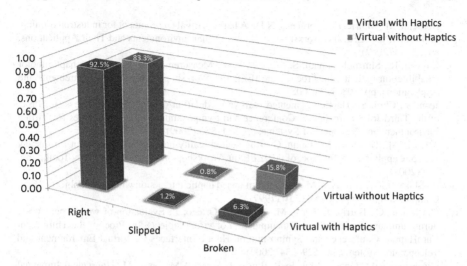

Fig. 9.5 Percentage of success tasks, broken and slipped objects for a Haptic and Visual collaborative manipulation task

revealed a high complexity in its development. For this kind of tasks, an architecture defined in this chapter has been required in order to properly manage all communication flows and software components. The MasterFinger-2 has been used effectively in simulating grasping and virtual collaborative tasks. It has demonstrated how two people can properly handle the same virtual object. This kind of simulation offers experts the opportunity to train in collaborative tasks. Potential applications for collaborative environments can be carried out in many fields such as training of manipulation skills (massage or rehabilitation), evaluation of assembly procedures, or research work about human behaviour in cooperative manipulation tasks.

Acknowledgements   This work was partly supported by the ImmerSence project within the 6th Framework Programme of the European Union, FET—Presence Initiative, contract number IST-2006-027141, see also www.immersence.info.

References

1. Kuschel, M., Buss, M., Freyberger, F., Farber, B., Klatzky, R.L.: Visual-haptic perception of compliance: fusion of visual and haptic information. In: Haptic Interfaces for Virtual Environment and Teleoperator Systems, pp. 79–86 (2008)
2. Turro, N., Khatib, O., Coste-Maniere, E.: Haptically augmented teleoperation. In: Proc. of the IEEE International Conference on Robotics and Automation, pp. 386–392 (2001)
3. Song, G., Guo, S., Wang, Q.: A tele-operation system based on haptic feedback. In: IEEE International Conference on Information Acquisition, pp. 1127–1131 (2006)
4. Acosta, E., Stephens, B., Temkin, B., Krummel, T.M., Gorman, P.J., Griswold, J.A., Deeb, S.A.: Development of a haptic virtual environment. In: Proc. of the 12th IEEE Symposium on Computer-Based Medical Systems, pp. 35–39 (1999)

5. Hosseini, M., Malric, F., Georganas, N.D.: A haptic virtual environment for industrial training. In: Proc. of the IEEE Int. Workshop on Haptic Virtual Environments and Their Applications, pp. 25–30 (2002)
6. Kayyali, R., Shirmohammadi, S., El Saddik, A.: Measurement of progress for haptic motor rehabilitation patients. In: Proc. of the IEEE Int. Workshop on Medical Measurements and Applications, pp. 108–113 (2008)
7. Jeon, S., Choi, S.: Haptic augmented reality: Modulation of real object stiffness. In: Proc. of the Third Joint EuroHaptics Conference and Symposium on Haptic Interfaces for Virtual Environment and Teleoperator Systems, pp. 384–385 (2009)
8. Adcock, M., Hutchins, M., Gunn, C.: Augmented reality haptics: using artoolkit for display of haptic applications. In: Proc. of the IEEE Int. Workshop on Augmented Reality Toolkit, pp. 1–2 (2003)
9. Salisbury, J.K., Srinivasan, M.A.: Phantom-based haptic interaction with virtual objects. IEEE Comput. Graph. Appl. **17**(5), 6–10 (1997)
10. Giachritsis, C., Barrio, J., Ferre, M., Wing, A., Ortego, J.: Evaluation of weight perception during unimanual and bimanual manipulation of virtual objects. In: Proc. of the Third Joint EuroHaptics Conference and Symposium on Haptic Interfaces for Virtual Environment and Teleoperator Systems, pp. 629–634 (2009)
11. Giachritsis, C.D., Garcia-Robledo, P., Barrio, J., Wing, A.M., Ferre, M.: Unimanual, bimanual and bilateral weight perception of virtual objects in the master finger 2 environment. In: Proc. of the 20th IEEE International Symposium on Robot and Human Interactive Communication (2010)
12. Hecht, D., Reiner, M., Halevy, G.: Multimodal virtual environments: response times, attention, and presence. Presence: Teleoperators and Virtual Environments **15**(5), 515–523 (2006)
13. Hecht, D., Reiner, M., Karni, A.: Enhancement of response times to bi- and tri-modal sensory stimuli during active movements. Exp. Brain Res. **185**(4), 655–665 (2008)
14. Iwata, H.: History of haptic interfaces. In: Grunwald, M. (ed.) Human Haptic Perception: Basics and Applications, pp. 355–363. Birkhäuser, Basel (2008)
15. Salisbury, K., Conti, F., Barbagli, F.: Haptic rendering: introductory concepts. IEEE Comput. Graph. Appl. **24**(2), 24–32 (2004)
16. Otaduy, M.A., Lin, M.C.: A modular haptic rendering algorithm for stable and transparent 6-dof manipulation. IEEE Trans. Robot. **22**(4), 751–762 (2006)
17. Mazzella, F., Montgomery, K., Latombe, J.C.: The forcegrid: a buffer structure for haptic interaction with virtual elastic objects. In: Proc. of the IEEE Int. Conf. on Robotics and Automation, pp. 939–946 (2002)
18. Payandeh, S., Azouz, N.: Finite elements, mass-spring-damper systems and haptic rendering. In: Proc. of the IEEE Int. Symposium on Computational Intelligence in Robotics and Automation, pp. 224–229 (2002)
19. Constantinescu, D., Salcudean, S.E., Croft, E.A.: Haptic rendering of rigid contacts using impulsive and penalty forces. IEEE Trans. Robot. **21**(3), 309–323 (2005)
20. Adams, R.J., Hannaford, B.: Stable haptic interaction with virtual environments. IEEE Trans. Robot. Autom. **15**(3), 465–474 (1999)
21. Ruspini, D., Khatib, O.: A framework for multi-contact multi-body dynamic simulation and haptic display. In: Proc. of the IEEE/RSJ International Conference on Intelligent Robots and Systems, pp. 1322–1327 (2000)
22. Garcia-Robledo, P., Ortego, J., Barrio, J., Galiana, I., Ferre, M., Aracil, R.: Multifinger haptic interface for bimanual manipulation of virtual objects. In: Proc. of the IEEE International Workshop on Haptic Audio visual Environments and Games, pp. 30–35 (2009)
23. Zilles, C.B., Salisbury, J.K.: A constraint-based god-object method for haptic display. In: Proc. of the 1995 IEEE/RSJ International Conference on Intelligent Robots and Systems, pp. 146–151 (1995)
24. Ruspini, D.C., Kolarov, K., Khatib, O.: Haptic interaction in virtual environments. In: Proc. of the 1997 IEEE/RSJ International Conference on Intelligent Robots and Systems, pp. 128–133 (1997)

25. Melder, N., Harwin, W.S.: Extending the friction cone algorithm for arbitrary polygon based haptic objects. In: Proc. of the 12th International Symposium on Haptic Interfaces for Virtual Environment and Teleoperator Systems, pp. 234–241 (2004)
26. Melder, N., Harwin, W.S.: Force shading and bump mapping using the friction cone algorithm. In: Proc. of the First Joint Eurohaptics Conference, 2005 and Symposium on Haptic Interfaces for Virtual Environment and Teleoperator Systems, pp. 573–575 (2005)
27. Wirz, R., Marin, R., Ferre, M., Barrio, J., Claver, J.M., Ortego, J.: Bidirectional transport protocol for teleoperated robots. IEEE Trans. Ind. Electron. **56**(9), 3772–3781 (2009)
28. Hannaford B, O.A.: Haptics. In: Siciliano, B., Khatib, O. (eds.) Springer Handbook of Robotics. Springer, Berlin (2008)

Chapter 10
Artificially Rendered Cutaneous Cues for a New Generation of Haptic Displays

Enzo Pasquale Scilingo, Matteo Bianchi, Nicola Vanello, Valentina Hartwig, Luigi Landini, and Antonio Bicchi

Abstract In this chapter we report on two architectures of haptic devices able to reproduce variable softness and elicit tactile sensations. Both solutions aim at addressing more effectively cutaneous channels. The first device is comprised of a tactile flow-based display coupled with a commercial kinesthetic interface. The second device is based on a pin array configuration in order to stimulate locally the fingertips and induce the illusion of different shapes or moving objects.

10.1 Introduction

Tactual sensory experience includes two distinct perceptual channels: kinesthetic information referring to the sensation of positions, velocities, forces and constraints arising from the muscle spindles and tendons. Force feedback haptic interfaces mostly rely on kinesthetic senses by presenting force-position control to create the illusion of contact with rigid or compliant surfaces. The cutaneous class of sensations arises through direct contact with the skin surface. When designing haptic displays, both of these channels should be elicited [1]. While it is quite simple to provide kinesthetic information, cutaneous information elicitation is still not sufficient. There are several attempts in literature to convey cutaneous information, e.g. by means of shape and/or vibration feedback, or by providing thermal data [2–4]. Here, we propose two technological solutions. The first one is based on the mechanical coupling of two displays while the second solution implies using a pin array display for locally deforming the fingertip skin.

The first display relies on a conjecture, first proposed by [5, 6], based on surrogating detailed tactile information for softness discrimination with information on the rate of spread of the contact area between the finger and the object as the contact

E.P. Scilingo (✉) · M. Bianchi · V. Hartwig · A. Bicchi
Interdepartmental Research Center "E. Piaggio", University of Pisa, via Diotisalvi, 2, 56126 Pisa, Italy
e-mail: e.scilingo@ing.unipi.it

N. Vanello · L. Landini
Department of Information Engineering, University of Pisa, via G. Caruso, 16, 56122 Pisa, Italy

A. Peer, C.D. Giachritsis (eds.), *Immersive Multimodal Interactive Presence*,
Springer Series on Touch and Haptic Systems,
DOI 10.1007/978-1-4471-2754-3_10, © Springer-Verlag London Limited 2012

force increases. This conjecture relies on the paradigm that a large part of haptic information necessary to discriminate softness of objects by touch is contained in the law that relates resultant contact force to the overall area of contact, or in other terms in the rate by which the contact area spreads over the finger surface as the finger is increasingly pressed on the object. Authors called this relationship Contact Area Spread Rate (CASR). Such a conjecture does not pretend to minimize the importance of other relevant aspects of tactile information, such as, e.g., shape of the contact zone or pressure distribution in the contact area, but it only suggests that the CASR information might increase tactile perception. For the sake of citation, next works [7] presented a softness display based on the control of fingertip contact area, although it was not able to display the dynamic change of the contact area and did not have enough spatial resolution.

Starting from a good resemblance between the growing rate of the contact area between the finger pad and an object during a tactile indentation task and the convergence or divergence of the vision field in time to contact task, we formulated a more general paradigm than CASR. In particular, the divergence from focus of expansion of optic flow represents the expansion of iso-brightness contours. The area delimited by a closed iso-brightness contour grows with motion over time likewise the growth of the contact area in the tactile domain. This analogy led us to define a new conjecture, inspired to optic flow, which we called *tactile flow*. Several works supported this conjecture [8, 9]. Hereinafter, we will refer to Tactile Flow (TF) display in place of CASR display.

The TF display, which mainly addresses cutaneous channels [10], is mechanically coupled with a commercial device, the Delta Haptic Device (DHD) by Force Dimension [11], which instead provides mostly kinesthetic information. In this chapter we will experimentally verify if this combination is able to constructively join the advantages of each device improving the overall performance. Indeed, the TF device is able to provide careful force/area and force displacement relationships, but due to mechanical constraints, these two behaviors are intimately related. On the other hand, the DHD can reliably provide force/displacement relationships, but it is not able to reproduce force/area relationships. Combining the two devices, therefore, we can independently control force/area and force/displacement relationships, thus extending the range of materials which can be reproduced. In addition, there are some materials which have the same force/displacement, but different force/area relationships (this latter, indeed, strongly depends on the geometry) and only an independent control of the two behaviors can replicate a similar rheology.

The second device is based on the concept of locally mechanically stimulating the fingertip skin in order to evoke different cutaneous sensations. The idea behind this device is to design a simple architecture which can be easily adapted to be used in fMRI environments. Generally, mechanical tactile displays utilize actuated components to actively deform the user skin via pressure, stretch or other means, in order to induce controlled touch sensations. They can be further classified by their method of stimulation into vibration, lateral displacement (skin stretch) and skin indentation. Vibration displays present shape information via activating patterns of spatially configured transducers at high temporal frequencies of operation.

These elements tend to be much smaller than the vibrotactile transducers, for example, pins or wires driven by solenoids or piezoelectric actuators [12]. Devices for lateral displacement present information through spatiotemporal patterns of skin stretch [13] and [14]. Tactile displays for skin indentation present distributed cues by inducing pressure on the skin via a number of moving elements. They have received the most attention for virtual environment applications as they offer the most significant potential to represent the fingertip deformations that occur in touch interaction with everyday objects. In literature there are several devices based on raised static pin patterns, as well as vibrotactile displays. The former find wide application in the development of embossed raised characters for visually impaired people, e.g. Braille display. Vibrotactile displays consist of a single element stimulator that is used to encode information in temporal parameters of the vibration signal. Parameters that can potentially be employed include frequency, amplitude, waveform, duration and rhythm [15]. Stimulation can be generated by various means including, but not limited to, solenoids, voice coils or rotation of an inertia by a motor. Most people make use of vibrotactile technology within mobile phones, as a non-audio-based indicator of an incoming call or text message. Vibrotactile information has also found widespread acceptance within the video gaming community as an inexpensive means of providing touch feedback in hand held controllers. Here we propose a mechanical display able to stimulate the fingertip with a pattern of pins and we investigate how tactile sensation changes with the frequency by which pins go up and down.

10.2 The Holistic TF-based Haptic System

The haptic system here proposed is comprised of a TF display [5] placed on the top of the Delta Haptic Device (DHD) (see Fig. 10.1). The TF display is a pneumatic device consisting of a set of cylinders of different radii in telescopic arrangement (see right side of Fig. 10.2). A regulated air pressure is inflated inside the display and acts on the cylinders so as to provide a simulated compliance which can be perceived by the operator when pushing with their forefinger against the top of the display. The schematic view of the TF display is shown on the left side of Fig. 10.2. A proportional Hall sensor placed at the bottom of the inner chamber allows measuring the displacement z of the cylinders when the subject pushes against them, while a servo pneumatic actuator regulates the chamber pressure according to the desired CASR profile to replicate. The Delta Haptic Device is a commercial interface, widely used by the haptic research community. It is a high performance haptic device and has 6 degrees of freedom: 3 translations from the parallel Delta structure and 3 rotations from a wrist module. Unlike other haptic mechanisms having either limited force capability or small workspace, the DHD is capable of providing large forces (up to 25 N) over a large volume (30 cm diameter, 30 cm length). In addition, because of its design and its base-mounted actuators, the device offers high stiffness, decoupled translation and rotation, as well as very low inertia. The haptic system exploits both performance of the two devices, joining the high fidelity and quality of tactile information of the TF display to the large workspace and high stiffness of the DHD.

Fig. 10.1 Picture showing
the haptic system

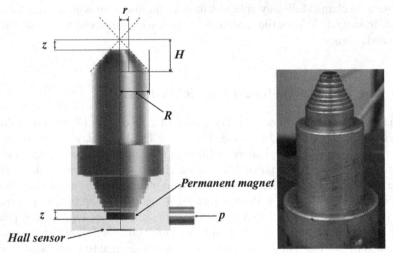

Fig. 10.2 Schematic view (*left*) and picture (*right*) of the TF display

10.3 Theoretical Justification

The initial contact of a mechanical interaction between two bodies may occur at a
point in case of spherical geometry or along a line in case of cylindrical bodies [16].
Let us focus on spherical geometry. Applying a slight load the area around the initial
contact point begins deforming. In such a way the mechanical interaction takes place
on a finite smaller area than bodies dimensions. Contact theory predicts the shape
of contact area and its behavior with time and with increasing load. Moreover, it

allows identifying stress and strain components in both bodies within and outside the loaded area.

When two spherical solids come into contact, the initial contact occurs at a single point which spreads over a circular area. According to the Hertz theory [16] the pressure exerted onto a generic circular area having radius r within the contact area between two solids of revolution is

$$p(r) = \frac{p_0}{a}\sqrt{(a^2 - r^2)}, \quad r \leq a$$

where a is the radius of the circular contact area and p_0 is the pressure at the origin. The total pressure is

$$P = \int_0^a p(r)2\pi r\,dr = \frac{2}{3}p_0\pi a^2.$$

Introducing the equivalent quantity

$$\frac{1}{E^*} = \frac{1 - v_1^2}{E_1} + \frac{1 - v_2^2}{E_2}$$

where E_1, E_2, v_1 and v_2 are the Young modulus and Poisson ratio for the two bodies coming into contact, respectively, we can obtain the radius of contact area:

$$a = \frac{\pi p_0 R}{2E^*}. \tag{10.1}$$

From this latter equation, it is worthwhile noting that Young modulus, contact area and curvature radius are strictly and intimately correlated. Young modulus relates the applied force (stress) with the induced displacement (strain). Equation (10.1) says that if two bodies are constitutively made of the same material, hence they have the same Young modulus, i.e. the same force/displacement, but different curvature radii, then they have different force/area behavior.

In order to replicate the rheology of these materials, it is necessary to implement an independent control of force/displacement and force/area. The TF display is able to replicate force/area behaviors with high reliability from a perceptual point of view, but it does not allow to implement two independent profiles of force/area and force/displacement. Indeed, when an external force is applied on the device it returns a reaction force given by $F = PA(z)$, where P is the pressure inflated into the inner chamber of the display and $A(z)$ is the contact area. This latter, however, is strictly related to the normal displacement by the bijective relationship $A(z) = \pi \frac{z^2 R^2}{H^2}$, where z is the normal displacement, R is the cylinder radius and H is the height of the cone (see Fig. 10.2). Therefore, when a force/area profile is set, force/displacement is indirectly obtained. On the other hand, DHD is a haptic interface which enables to reliably replicate force/displacement curves, but it does not allow to provide force/area behavior. By coupling the two devices, it is possible to control independently force/area and force/displacement, joining synergistically the two performance.

10.4 Hardware Equipment: Modeling and Identification

In this section, we modeled the TF device in order to analytically characterize it in terms of force/area relationships parameterized with by pressure. Afterwards, these curves were experimentally validated inflating inside the TF device progressive constant pressures and measuring the force/area profile at each pressure level.

10.4.1 Theoretical Model of TF Display

If we assume the system lossless, while the probing finger pushes with a constant force F against the top of the display, this latter exerts an equal and opposite reaction at the equilibrium position in agreement with the virtual work principle:

$$F\delta z = -p\delta V \quad \Rightarrow \quad F = -p\frac{\delta V}{\delta z},$$ (10.2)

where δz and δV are respectively the virtual displacement along the z axis, and the virtual variation of the volume V of the TF display. The variation of the inner volume of the TF display is given by

$$\delta V = -\pi \frac{z^3 R^2}{3H^2},$$

in which $H = 0.01$ m is the maximum height, and $R = 0.0065$ m is the radius calculated at the basis of the CASR display. According to (10.2), we easily obtain the analytical model for the display

$$F = p\pi \frac{z^2 R^2}{H^2} = pA.$$ (10.3)

10.4.2 Experimental Model of TF Display

The analytical model previously calculated has been experimentally assessed using the hardware setup shown in Fig. 10.3. The TF display was submitted to indentation tests at different pressures by means of a compressional indentor driven by an electromagnetic mini-shaker. The actuator, made by Bruel & Kjear, is a linear current step motor mini-shaker type, capable of applying a maximum displacement of 10 mm in axial direction. The indentor is a metallic cylinder of 1.5 cm in diameter and 10 cm in length. The indentor is equipped with a magnetic linear transducer, Vit KD 2300/6C by KAMAN Science Corporation, used to measure the applied axial displacement and with a load cell sensor, ELH-TC15/100 by Entran, able to detect forces up to ±50 N.

Fig. 10.3 Experimental
setup used for identifying the
TF model

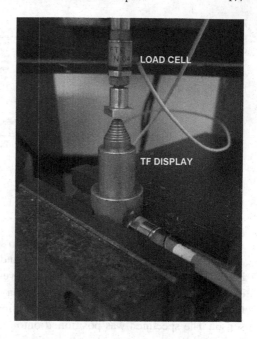

An external electronic driver is used to activate the indentor and acquire force-position signals from the sensors.

A data acquisition PCI card with one analog output channel and two analog input channels, is used to gather signals and sent them to a PC. A dedicated software was implemented in Matlab/Simulink environment to control in feedback and in real-time the displacement of the indentor and to record and plot the signals.

10.4.3 Identification Assessment

The TF display was characterized in terms of force versus axial displacement at different values of pressure. Keeping the pressure constant, as the shaker pushes onto the TF display the outputs from both force and position sensor are recorded. The pressure inside the display is maintained constant as the shaker pushes by an internal control of a servovalve, Proportion-Air's QB series, employed for this purpose. The experimental curves were compared with the theoretical ones. In Fig. 10.4 three experimental force/area curves at three different values of pressure were compared with the corresponding theoretical ones. Results show a satisfactory agreement between three any TF force/displacement curves and those theoretically calculated at the same pressures. This supports the hypothesis that the air loss inside the TF display can be considered negligible.

178

E.P. Scilingo et al.

Fig. 10.4 Force/Displacement response of the TF display compared with the analytical model at three different levels of pressure, by way of illustration

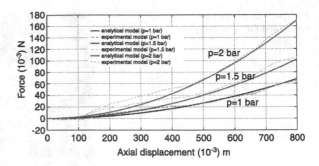

10.5 Experimental Session

An experimental protocol was applied to two sphere-shaped viscoelastic specimens, M_1 and M_2, approximately homogeneous and with different curvature radii, where both force-area and force-displacement curves are acquired. Figure 10.5 shows the experimental setup used to acquire force/area curves. The same hardware setup used for characterizing the TF display was additionally equipped with a dedicated area sensor. The specimen was positioned onto a transparent plexiglass surface whereunder a web camera was placed. As the indentor pushes against the specimen the web cam captures a snapshot of the surface flattened against the plexiglass. In order to enhance contours of contact area a thin white paper behaving as optical filter was placed between the specimen and the plexiglass. In addition to force/area, also the indentation depth is detected by a magnetic position sensor. In Fig. 10.6 two snapshots of the contact area were captured on the same specimen at two different level of indenting force are shown. The experimental force/area curves acquired by the hardware setup previously described as well as the corresponding mathematical interpolated curves are reported in Fig. 10.7 for the two specimens. A quadratic interpolation provided best fitting.

Two spherical specimens of the same material M_1, having radii $R = 1.1$ cm and $R = 1.3$ cm, respectively, and two specimens of the material M_2 having radii $R = 1.4$ cm and $R = 1.7$ cm were characterized in terms of force/area and force/displacement. Experimental curves of force/area were interpolated and as expected from the Hertz theory best fitting was obtained by a quadratic interpolation. These equation were used in the control law. Here we report, by way of illustration, the interpolated equation for the specimen of radius $R = 1.1$ cm of the material M_1:

$$F(A)|_{(M_1, R=1.1)} = 1.4A^2 + 0.33A + 0.55. \tag{10.4}$$

Figure 10.7 shows all the four experimental force/area relationships along with the interpolated ones, for the material M1 and M2, respectively on the left and right. An analogous procedure was done for force/displacement identification. Results are reported in Fig. 10.8 left and right side. As can be seen, these curves are mathematically approximated by a linear interpolation.

Fig. 10.5 Experimental setup used for identifying force/area curve of the specimens

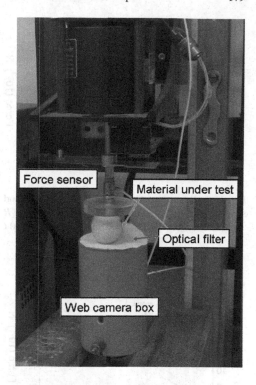

Fig. 10.6 Two different snapshots of the contact area captured on the same specimen having radius $R = 1.3$ cm at two level of force

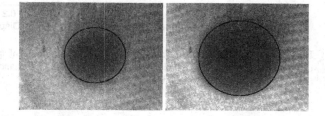

For the sake of brevity, here we report the equation of the interpolated curve for the specimen of material M_2 with radius $R = 1.4$ cm

$$F(z)|_{(M_2, R=1.4)} = 23z + 0.97. \tag{10.5}$$

It is worthwhile noting that the slope of the interpolated curves for the same material is the same, in agreement with the theory that same materials, though with different geometry, have the same Young modulus, hence roughly the same force/displacement curves.

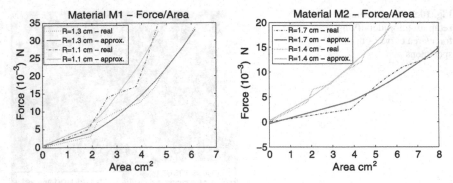

Fig. 10.7 Experimental force/area relationships and interpolated curves for two sphere-shaped specimens of the same material M_1 having radius ($R = 1.3$ cm and $R = 1.1$ cm) on the left side and M_2 having radius ($R = 1.7$ cm and $R = 1.4$ cm) on the right side

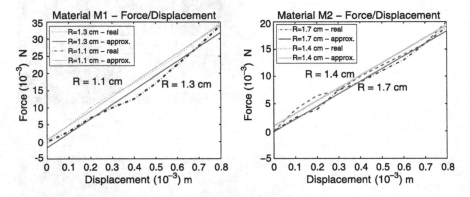

Fig. 10.8 Force/Displacement relationships and approximations of spherical specimen M_1 having different radius ($R = 1.3$ cm and $R = 1.1$ cm) on the left side and M_2 having different radius ($R = 1.7$ cm and $R = 1.4$ cm) on the right side

10.5.1 Control Strategy

The proposed, integrated haptic system allows replicating any force/area and force/displacement relationships, thanks to the possibility of independently controlling the displacement z_D of the DHD in addition to the air pressure p inflated inside the CASR device. If we denote, indeed, z_C and z_D the displacements of the CASR and of the DHD respectively, both force/area and force/displacement relationships can be mapped by the independent set of controls p and $z_D = z_m - z_C$, where z_m is the material displacement to replicate.

Let us suppose to replicate, e.g., the material M_1, neglecting the terms of zero and first order in the force/area equation as well as the intercept in the straight-line equation of force/displacement:

$$\begin{cases} F_m = \alpha_m A_m^2, \\ z_m = \beta_m F_m. \end{cases} \tag{10.6}$$

The force/area relationship should be reproduced by the CASR device, whose force/area characteristic is reported in (10.3). Imposing the same force and area of the material to replicate, we get

$$p = \alpha_m A_m. \tag{10.7}$$

In order to track the force/displacement behavior, we should set

$$z_m = z_C + z_D, \tag{10.8}$$

where z_C is read by the position sensor placed inside the CASR display and is analytically given by the combination of (10.3) and (10.7).

Coupling the two devices mechanically in series, we have $F_C = F_D$ which in turn should be equal to F_m. This force is the input of DHD, while the output z_D is controlled in feedback so that $z_D = z_m - z_C$, which can be expressed as

$$z_D = z_m - z_C = \beta_m F_m - z_C = \beta_m p \frac{\pi R^2}{H^2} z_C^2 - z_C. \tag{10.9}$$

10.6 Experimental Results

In order to assess the performance of the control strategy, we replicated the behavior of the materials M_1 and M_2, previously identified. Pushing on the haptic system with a desired force profile, the control law responded in terms of area and displacement in agreement with the analytical model of materials previously identified. Output from displacement and area sensors of the haptic system were compared with the analytical desired trajectories and we reported experimental results in Figs. 10.9 and 10.10. As it can be seen, this new architecture is able to successfully track the theoretical curves with negligible errors. In addition to good tracking performance, the combined system is also able to provide increased haptic feeling and further psychophysical experiments aiming at assessing that are planned to be performed. This confidence is supported by results from previous works [5]. Indeed, in the new combined system here proposed, cutaneous information is mostly provided by the CASR display which was already shown to better discriminate softness with respect to a purely kinaesthetic display but additionally kinaesthetic information is increasingly enhanced by DHD performance, in terms of accuracy and reliability.

10.7 The Pin-Based Mechanical Stimulator

One approach to render spatial aspects of the tactile sense is the use of pin arrays, i.e. a fingertip sized array of small pins that can be moved individually. Movement can

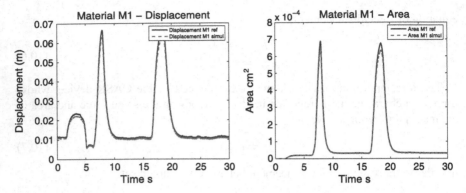

Fig. 10.9 Tracking of the displacement (on the left side) and area (on the right side) of M_1 by the haptic system. *Continuous line* represents the response of the model of the material to an external force, while *dashed line* is the tracking output of the system

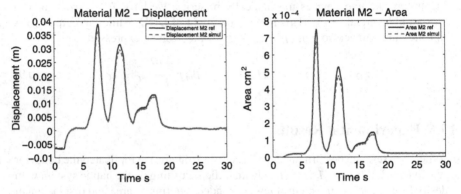

Fig. 10.10 Tracking of the displacement (on the left side) and area (on the right side) of M_2 by the haptic system. *Continuous line* represents the response of the model of the material to an external force, while *dashed line* is the tracking output of the system

be modulated in frequency up to achieve a vibratory stimulation. Here we propose a very simple haptic display which is based on a 4×4 pin array configuration. This architecture was thought to be fMRI compatible. This haptic display is comprised of a 4×4 pin array, which is moved by 16 servo motors. Each motor is controlled independently using a graphic interface in LabView. It is possible to change the frequency of the control signal hence the stimulator pin velocity. Pins are constituted of 1 mm diameter metal bars with a round rubber tip to avoid any finger injuries and are spaced 2 mm apart. Sixteen servo motors are used to move the pins up and down, in vertical direction: these motors have a plastic case and are controlled by a pulse signal. The rotation angle of the motor axes is correlated to the pulse width, so that, by changing the parameters of the control signal, it is possible to vary the movement of the pin bar. All the motors are fixed on a dedicated PVC structure with a perforated ceil to support the pin array (see Fig. 10.11). Motor control is obtained using a DAQ National Instrument acquisition card and a notebook with LabView

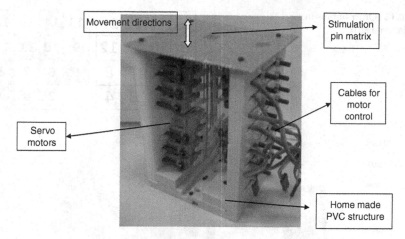

Fig. 10.11 Mechanical structure of the device

software. Using a graphic interface it is possible to choose the parameters of the control signals of each motor: duty cycle, frequency, pulse width, signal length and phase. The amplitude of all control signals is 1 V. Through the graphic interface the user can choose the pin to act. Each of the 16 pins can be actuated independently or combined.

10.7.1 Experimental Tests

Preliminary tests on the new haptic device were performed in order to assess performance. Two experimental sessions were set out. The first one aimed at investigating how frequency affects the capability of localizing the mechanical stimulating point. The second experiment focused on a preliminary study for verifying whether the device is able to induce illusionary movements and therefore to replicate the paradigm of tactile flow by stimulating the pins in the manner of a progressive wave.

Spatial Localization

The first test focused on the identification and localization of a tactile stimulus obtained with one pin. Subjects were asked to place their right forefinger on the stimulation array and lean their wrists on a support to limit hand movements. Only one of the 16 pins was actuated using different frequencies (3, 5, 10 Hz): the subject was asked to indicate the number of the moving pin according to the scheme shown in Fig. 10.12, left side. The test involved 5 stimulation sessions: for each session only one pin was moved for 10 seconds with three different stimulation frequencies (3, 5 10 Hz) using a pulse train with 5 pulse of 1.5 ms and five pulses of 2 ms. At the end

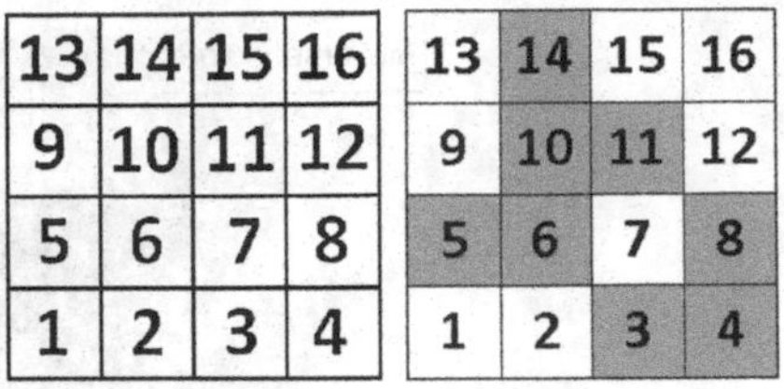

Fig. 10.12 Order of pins in stimulation matrix

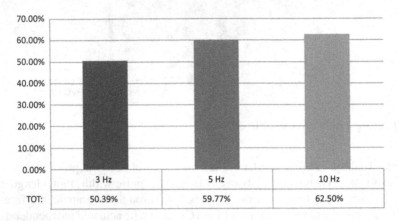

Fig. 10.13 Correct localization percentage

of each stimulation session subjects were asked to indicate the pin activated reporting the number according to the matrix scheme. Due to technical limitations only 8 pins out of 16 were used and more specifically the numbers indicated in Fig. 10.12, right side, but subjects were in the dark about that. The different pins were activated with different frequencies according to a random protocol. Eight subjects participated in the test (5 male, 3 female, age 25–30 years, right handed). Figure 10.13 shows the results of the described test in terms of correct localization percentages, for each stimulation frequency used. It is worthwhile noting that most of exact responses (62.5%) were obtained at the highest frequency (10 Hz) and the number of exact responses increases as the stimulation frequency grows. This can be interpreted as a wider involvement of SAI mechanoreceptors (Merkel corpuscles) and RA afferents (Meissner's corpuscles) which are most sensitive to this dynamic mechanical stimulus at that frequency. Moreover, they are the most numerous receptor types with fine spatial sensitivity and are best at the fingertips. In addition, the analysis of the results indicates that, independently of the stimulation frequency used, the best localized pin is the number 4. This would imply that subjects are more able to localize stimulating point at the edges than central points. In order to complete the study, we defined four types of error:

Error A: The subject indicates a pin right adjacent to the activated one on the same row or the same column. For example: pin stimulated = 10, pin indicated by the subject = 9 or 11 or 14 or 6;

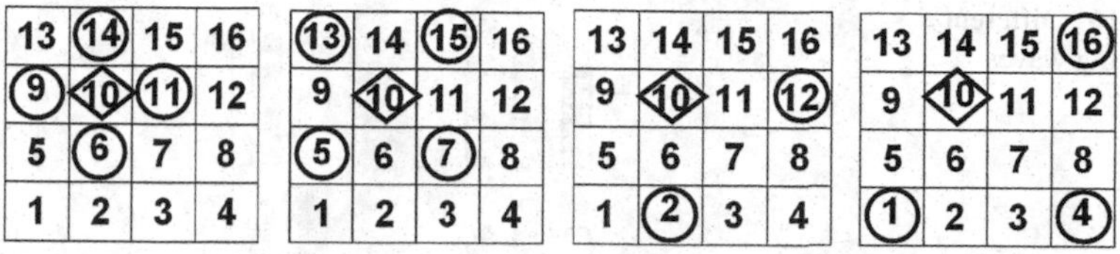

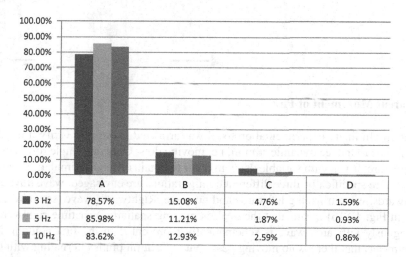

Fig. 10.14 Types of error in localizing the stimulating point, from A to D reading from left to right

	A	B	C	D
■ 3 Hz	78.57%	15.08%	4.76%	1.59%
■ 5 Hz	85.98%	11.21%	1.87%	0.93%
■ 10 Hz	83.62%	12.93%	2.59%	0.86%

Fig. 10.15 Percentage of errors made by the subjects

Error B: The subject indicates a pin close to the activated one but not on the same row or same column. For example: pin stimulated = 10, pin indicated to the subject = 13 or 15 or 5 or 7;

Error C: The subject indicates a pin on the same row or the same column of the activated pin but not close to this one. For example: pin stimulated = 10, pin indicated to the subject = 2 or 12;

Error D: The subject indicates a pin not on the same row or the same column of the activated pin and not close to this one. For example: pin stimulated = 10, pin indicated to the subject = 16 or 1 or 4.

Figure 10.14 shows in graphical form the four types of errors. Figure 10.15 shows the percentages of the four error types at different stimulation frequencies. For all the frequencies used the most frequently occurred is the A type. This means that, even if subjects did not identify correctly the stimulation point, she/he can approximately localize the right stimulation position in the most of cases.

Fig. 10.16 Different
stimulating waves

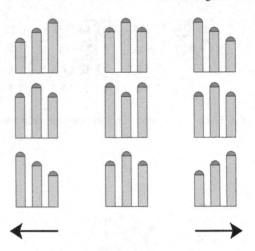

Apparent Movement of Pins

A very preliminary experimental protocol was arranged in order to verify if the device is able to induce dynamic stimuli, i.e. moving dots. Only one single row is used and pins are driven with variable delay in order to obtain different stimulation conditions. More specifically, three different configurations are envisaged: wave traveling rightwards, wave traveling leftwards and no wave. Rightwards wave (see third column in Fig. 10.16) means that the peak is moving spatially over time rightwards; analogously leftwards wave is schematically represented in the first column. No wave means that there is no moving peak, but the central pin is alternating with the two lateral ones. A group of 21 subjects volunteered to participate in the experiment, 8 males and 134 females, age comprised between 21 and 24, all right-handed. The experiment consist in two trials of six stimulations, and each configuration is presented randomly twice. The aim was to investigate the capability of recognizing the movement of the virtual wave and possibly the direction. Two types of analysis were carried out. The first one aimed at assessing whether subjects are able to perceive a moving dot under the fingerpad. The second aimed at analyzing whether subjects that perceived the movement are also able to correctly identify the direction of motion. Results are reported in Fig. 10.17.

10.8 Conclusion

In the first part of this chapter we proposed a new configuration of a haptic system comprised of two devices mechanically coupled in series. This new architecture allowed to implement an independent control of force/area and force displacement, extending the range of materials which can be replicated. After introducing the theoretical motivation we identified four specimens consisting of two materials, M_1 and M_2 each one of two different sizes. The two specimens having different geometry of

Fig. 10.17 77% of subjects perceived a moving pin under the fingerpad and among these 65% identified correctly the direction

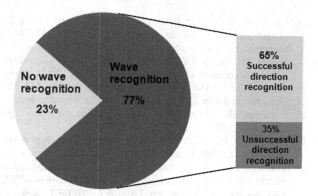

the same material M_1 exhibit the same force/displacement, but different force/area. They can be replicated only by a device able to implement an independent control of the two relationships. The same is true of the material M_2. Experimental results showed that the newly proposed system is able to effectively track every curve of each material. The last part of the chapter was dedicated to evaluating performance of a pin array device. The idea behind was to verify if the device is able to reproduce simple shapes using several pins simultaneously and illusionary moving patterns. Finally, future work will address the compatibility of the new haptic device for fMRI based investigations of brain functions related to tactile tasks, in order to verify the hypothesis of the recruitment of different areas by changing stimulation parameters.

Acknowledgements   This work was partly supported by the ImmerSence project within the 6th Framework Programme of the European Union, FET—Presence Initiative, contract number IST-2006-027141, see also www.immersence.info.

References

1. Caldwell, D.G., Lawther, S., Wardle, A.: Multi-modal cutaneous tactile feedback. In: IEEE/RSJ International Conference on Intelligent Robots and Systems IROS, Osaka, Japan, pp. 465–472 (1996)
2. Hayward, V., Astley, O.R., Cruz-Hernandez, M., Grant, D., Robles-De-La-Torre, G.: Haptic interfaces and devices. Sens. Rev. **24**(1), 16–29 (2004)
3. Yano, H., Komine, K., Iwata, H.: Development of a high-resolution surface type haptic interface for rigidity distribution rendering. In: First World Haptics Conference, Pisa, Italy, pp. 465–472 (2005)
4. Srinivasan, M.A., LaMotte, R.H.: Tactile discrimination of softness. J. Neurophysiol. **73**(1), 88–101 (1995)
5. Bicchi, A., De Rossi, D.E., Scilingo, E.P.: The role of the contact area spread rate in haptic discrimination of softness. IEEE Trans. Robot. Autom. **16**(5), 496–504 (2000)
6. Bicchi, A., Scilingo, E.P., Dente, D., Sgambelluri, N.: Tactile flow and haptic discrimination of softness. In: Barth, F.G., Humphrey, J.A., Secomb, T.W. (eds.) Multi-point interaction with real and virtual objects. Springer Tracts in Advanced Robotics, pp. 165–176 (2005)
7. Fujita, K., Ohmori, H.: A new softness display interface by dynamic fingertip contact area control. In: 5th World Multiconference on Systemics, Cybernetics and Informatics, Pisa, Italy, pp. 78–82 (2001)

8. Bicchi, A., Scilingo, E.P., Ricciardi, E., Pietrini, P.: Tactile flow explains haptic counterparts of common visual illusions. Brain Res. Bull. **75**(6), 737–741 (2008)
9. Ricciardi, E., Vanello, N., Sani, L., Gentili, C., Scilingo, E.P., Landini, L., Guazzelli, M., Bicchi, A., Haxby, J.V., Pietrini, P.: The effect of visual experience on the development of functional architecture in hMT+. Cereb. Cortex **17**(12), 2933–2939 (2007)
10. Goldstein, E.B.: The cutaneous senses. Sens. Percept. **443**, 329–355 (2002)
11. Delta haptic device (2011). http://www.forcedimension.com
12. Summers, I.R., Chanter, C.M.: A broadband tactile array on the fingertip. J. Acoust. Soc. Am. **112**, 2118 (2002)
13. Pasquero, J., Hayward, V.: Stress: A practical tactile display system with one millimeter spatial resolution and 700 Hz refresh rate. In: Proc. Eurohaptics, vol. 2003, pp. 94–110 (2003)
14. Caldwell, D.G., Tsagarakis, N., Giesler, C.: An integrated tactile/shear feedback array for stimulation of finger mechanoreceptor. In: Proceedings of IEEE International Conference on Robotics and Automation, vol. 1, pp. 287–292. IEEE, New York (1999)
15. Brewster, S., Brown, L.M.: Tactons: structured tactile messages for non-visual information display. In: Proceedings of the Fifth Conference on Australasian User Interface, vol. 28, pp. 15–23. Australian Computer Society, Perth (2004)
16. Johnson, K.L.: Contact Mechanics. Cambridge University Press, Cambridge (1985)

Chapter 11
Social Haptic Interaction with Virtual Characters

Zheng Wang, Jens Hölldampf, Angelika Peer, and Martin Buss

Abstract Adding physicality to virtual environments is considered a prerequisite to achieve natural interaction behavior. While physical properties and laws can be built into virtual environments by means of physical engines, providing haptic feedback to the user requires appropriately designed and controlled haptic devices, as well as sophisticated haptic rendering algorithms. While in the past a variety of haptic rendering algorithms for the simulation of human-object interactions were developed, haptic interactions with a virtual character are still underinvestigated. Such kind of interactions, however, pose a number of new challenges compared to the rendering of human-object interactions as the human expects to interact with a character that shows human-like behavior, i.e., it should be able to estimate human intentions, to communicate intentions, and to adapt its behavior to its partner. On this account, algorithms for intention recognition, interactive path planning, control, and adaptation are required when implementing such interactive characters. In this chapter two different approaches for the design of interactive behavior are reviewed, an engineering-driven and a human-centred approach. Following the latter approach virtual haptic interaction partners are realized following the workflow record-replay-recreate. To demonstrate the validity of this approach it is applied to two prototypical application scenarios, handshaking and dancing.

Z. Wang (✉) · J. Hölldampf · A. Peer · M. Buss
Institute of Automatic Control Engineering, Technische Universität München, 80290 München, Germany
e-mail: wang@tum.de

J. Hölldampf
e-mail: jens.hoelldampf@tum.de

A. Peer
e-mail: angelika.peer@tum.de

M. Buss
e-mail: mb@tum.de

A. Peer, C.D. Giachritsis (eds.), *Immersive Multimodal Interactive Presence*,
Springer Series on Touch and Haptic Systems,
DOI 10.1007/978-1-4471-2754-3_11, © Springer-Verlag London Limited 2012

11.1 Introduction

While today's virtual environments encountered in applications like simulation, training, rehearsal, and virtual gatherings are able to provide high quality visual and auditory feedback, most of them still lack physicality, the state or quality of being physical and follow physical principles: People can penetrate into walls, objects, and characters, lift objects without feeling their weight, stroke objects without feeling their texture and socially interact with characters, without feeling forces when being in physical contact.

One of the main challenges in designing virtual environments, however, is to provoke natural interaction behavior. Among others, physicality is considered one of the main prerequisites to achieve this. While physical engines are typically used to build physical properties and laws into virtual environments, providing additional haptic feedback calls for appropriately designed and controlled haptic devices, as well as sophisticated haptic rendering algorithms.

In recent years a variety of haptic devices have been developed and presented, see [1, 2] for an overview. Unfortunately, most of them have been designed and optimized for a specific application only and thus, establishing virtual environments with the capability of providing general haptic feedback is still considered very challenging.

Beside haptic devices, also high-quality haptic rendering algorithms contribute to achieving physicality of virtual environments. Haptic rendering has been a very active field of research and a variety of algorithms for human-object interaction have been developed in the past. This includes geometric rendering algorithms for single point contact with polygonal [3, 4], parametric [5], and implicit surfaces [6] or volumetric objects. Advanced versions also consider point interaction with deformable objects, line interaction or interaction between polygons. Recently, also direct haptic rendering from measurements has been studied intensively [7]. Beside rendering algorithms for kinesthetic feedback also a number of texture rendering algorithms exist, see [8, 9]. Interested readers should refer to [10] for a comprehensive overview of state-of-the-art haptic rendering algorithms.

While all the aforementioned rendering algorithms focus on human-object interactions only, the rendering of haptic human-human interactions have been rarely studied in literature. Thus, in this work we focus on the rendering of a virtual, interactive character that can haptically interact with a human partner in a social interaction task like handshaking or dancing. Compared to haptic rendering of human-object interactions, rendering of a virtual, interactive character that is able to perform social haptic interaction tasks poses a variety of new challenges as the human expects to interact with a character that shows human-like behavior, i.e., it should be able to estimate human intentions, to communicate intentions, and to adapt its behavior to its partner. Thus, in contrast to the rendering of human-object interactions, the rendering of plausible interactive characters requires additional modules for intention recognition, interactive path planning, control, and adaptation.

First approaches implementing such modules can be found in the field of physical human-robot interaction where a robot is supposed to physically assist a human operator while performing a joint transportation or social interaction task. Starting from purely passive followers as presented in [11], varying impedance parameters [12–17] and controllers that introduce additional virtual constraints [18] were developed, leading finally to active robot partners that can estimate human intentions [19–21] and based on these estimations change their interaction behavior by taking over different roles [22, 23].

In the following section, we will present our approach adopted for realizing haptically interacting characters, while Sects. 11.3 and 11.4 exemplarily demonstrate it for two prototypical social interaction tasks, handshaking and dancing.

11.2 Approaches to Synthesize Interactive Behavior

When aiming for the design of an interactive character, two completely different approaches can be adopted: an engineering-driven and a human-centered approach. A typical engineering-driven approach would implement control strategies based on e.g. optimality or stability criteria. One of the drawbacks of such an approach is that it does not necessarily guarantee that the resulting interaction patterns are human-like which complicates the recognition of the virtual partner's intention, the building of its mental model and thus prediction of its action. Beside this, natural communication of one's own intentions can be negatively affected by the usage of such an approach, as the artificial virtual partner often lacks the ability to understand and interpret them.

These limitations can be overcome by a human-centered approach which aims at first studying human–human interactions, then analyzing observed behavioral patterns, building computational models, and finally, transferring obtained knowledge and models to human-robot interaction [21, 24–32].

In the EU project ImmerSence,[1] we adopted this human-centered approach to synthesize interactive behavior following a *record-replay-recreate* workflow. In the first recording phase force and motion signals resulting from haptic interaction of two humans were recorded. In the second phase, this data was simply replayed by using a haptic interface. Since, a pure replay lacks the ability to adapt to the human partner and thus natural interaction behavior could not be achieved [26], a third phase, the recreation phase was introduced which aimed at synthesizing real interactive behavior. The following two sections will demonstrate this approach for two prototypical application scenarios, handshaking and dancing.

11.3 Application Scenario: Handshaking

In many cultures around the globe, people greet each other by handshaking. Yet little is known on how people plan or carry out such a process, or whether a robot can

---

[1]www.immersence.info, 2011.

Fig. 11.1 Handshaking with a virtual interactive character

replace one human in a handshake while giving the other human the illusion of a realistic and plausible handshake. In this work, we follow the human-centered *record-replay-recreate* approach to imitate human-like handshakes with a robot. Starting from the *recording* of human-human handshakes, the robot is first programmed to imitate human behavior by trying to *replay* the recorded handshakes. Then, since a pure replay lacks the ability of adaptation to the partner, *recreation* of handshakes based on a dynamical model is investigated.

So far, only few authors investigated handshaking: in [33] the first tele-handshake using a simple one degree-of-freedom (DOF) device was created while [34] generated handshake animations from a vision system. Remarkably, only few people viewed handshaking from a force/motion interaction aspect: in [35] the authors took the oscillation synchronization approach to realize human-robot handshaking and in [36] the authors focused on the approaching and shaking motions of a handshaking robot.

In this work, we view handshaking as a haptic interaction process, and successfully recreate the entire handshaking process between human participants and our robotic handshake partner. In order to achieve this, we developed a virtual reality system that consists of haptic, vision, and sound components, see Fig. 11.1. Here the discussion is focused only on the haptic component.

An approach to realize a virtual handshake partner is to adopt algorithms originating from imitation learning or learning from demonstration, which is a well established research field. Main research in this area, however, has been devoted to the learning of motion skills, see [37] and [38]. Only recently researchers started to incorporate force information, see e.g. [39] for the learning of a door-opening and an ironing task and [40] for a grasping task. Interaction with passive objects, however, as studied in these examples, differs significantly from handshaking, because in a handshaking process the partners mutually influence each other and adapt to each other. Consequently, interaction skills cannot be treated like motion skills and thus, classical imitation learning methods need to be adapted and extended.

Another approach known from literature is to first estimate human intentions and then to change the robot behavior accordingly. In [20] for example, a female dancing robot was developed with the ability to estimate human dancing steps, while in [41] the author investigated robotic assistance in a calligraphy task. In both applications, different robot behavior was implemented for all of the estimated human intentions. In this work we will follow a similar approach by differentiating first between two different human interaction strategies in handshaking and then by assigning the robot the respective opposite role by activating respective behavioral models, see [31] and [42] for our first presentations of this idea.

Independently from our group, the authors in [22, 43] developed a similar approach for a joint lifting task. Again, different roles of the interacting partner were distinguished and the respectively opposite role was assigned to the robot. But instead of activating predefined robot behavioral models, the required robot behavior for each of the roles was first learned from highly controlled human demonstrations using Hidden Markov Models or Gaussian Mixture Models and then reproduced on demand. The authors estimated roles from interaction force and motion directly, while we will show that human behavioral parameters are the better selection as they allow to remove the robot influence on the signals and to distinguish between an actively resisting (active) human and a (passive) human that acts like an inertia and passively resists the robot motions.

In the following sections, we will review our approach taken for the development of an interactive handshaking partner following the *record-replay-recreate* workflow delineated above.

11.3.1 Recording

In the recording phase, participants were asked to perform handshakes with each other in the same way they would do in their normal lives. The aim was to gather first-hand data of human-human handshake processes to serve as a reference for later studies. Position and force were measured for each handshake for robotic control.

In the handshake recording experiment a total number of 900 human-human handshakes of 24 male college students (mean age 28, standard deviation 2.292) were performed, where position and force data was recorded during interaction by special data gloves [44] and an optical tracking system, see Fig. 11.2. No instructions on how to perform the handshakes were given to achieve natural interaction behavior.

The optical tracking system used in the recording provided 3-dimensional measurements of handshake trajectories, see Fig. 11.3. The following findings were drawn from the recorded handshakes:

- Handshake is a dynamic process with intra- and inter-subject variations. The actual trajectory is affected by both sides of the handshake pair.
- A typical handshake can be divided into three stages: approach, shake, and release. The switching conditions are the contact and lose of contact between the two hands of the participants.
- In the shaking stage of a handshake, force and position are exchanged among partners, while the other two stages involve only free space motion.

Fig. 11.2 Equipments used
for recording human-human
handshakes. The left
participant wears the data
glove to measure gripping
forces, while optical motion
tracking systems are placed in
the background. The right
person is only included for
illustration purposes. In
actual recordings both
participants wore identical
equipments for recording

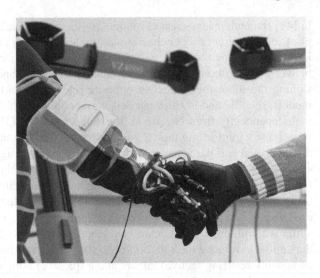

Fig. 11.3 Wrist trajectory of
one participant during 30
handshakes with same
partner. The lower part of the
trajectory (close to 750 on the
vertical axis) is the place
where the participant's hand
rests between handshakes,
while the higher part (close to
1100 on the *vertical axis*)
shows the area where
handshakes take place

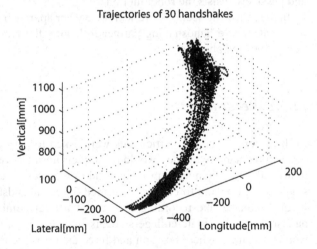

- Large inter-trial similarities between subjects can be observed for the trajectories of reaching and retreating stages, while the shaking stage differs significantly.

Based on these observations, first a replay-version of handshakes was realized, followed by the recreation of them based on a dynamical system.

11.3.2 Replay—A Leading Partner Implementation

In the replay phase trajectories recorded during human-human handshakes were replayed using a haptic interface. The robot was controlled with a position controller that follows a desired trajectory. The desired trajectory was pre-recorded from human-human experiments. Assuming the robot has adequate actuation power, the actual position trajectory will be the same as the reference, regardless of the action of the human partner. As interaction forces from the human partner are not considered, this implementation represents a non-compromising partner, or in other words, a partner with no compliance.

If the human performs similarly at each handshake, the same desired trajectory leads to similar force profiles for each handshake and hence provides the human a similar handshaking experience compared to the human-human case where the reference trajectory was recorded. Thus, provided that human participants are good repetitors, this replay strategy leads to natural handshakes. If, however, the human desires to take over the lead, the compliant controller's disability to produce compliance and to adapt to the human partner results in a rather unrealistic and unnatural handshake experience.

11.3.3 Recreate—An Interactive Partner Implementation

Thus, an advanced interactive handshaking controller was implemented that aimed at imitating real human handshaking behavior. The term *recreate* is used to stress the fact that handshakes are generated model-based adopting knowledge extracted from pre-recorded human-human handshakes, but that resulting handshakes are generally different from any pre-recorded data.

Based on the observations made from human-human recordings, the handshake process was first segmented into three stages: approaching, shaking, and retreating. A three state switching model was constructed corresponding to the three handshake stages. The similar patterns of approaching and retreating stages were fitted by bell-shaped curves from recorded data for the robot to follow in free-space motion. In the following subsection we will concentrate on the shaking stage only.

Basic Handshake Model: Compliant Partner

In [45] we presented a first handshake prototype that imitates the human arm behavior during handshaking. Robot behavior was modeled by a second order impedance model as shown in (11.1), where M, B, and K are the impedance parameters mass, damping, and stiffness representing the robot arm, f and x denote force and position, and x_0 is the equilibrium position of the virtual partner arm:

$$f(t) = M\ddot{x}(t) + B\dot{x}(t) + K(x(t) - x_0). \tag{11.1}$$

While for the arm mass and damping constant parameters were chosen, the stiffness parameter was adapted to imitate human arm stiffness and to improve naturalness of interaction. In practice, the contraction of the muscle groups is increased when the human is exerting higher forces. The stiffness of the arm is therefore increased. Inspired by this fact, a time-varying virtual stiffness K consisting of a constant term K_0 in addition to a term proportional to the difference between the actual position $x(t)$ and the equilibrium position x_0, defined as the neutral position of the robot end-effector, was implemented:

$$K = K_0 + \eta(x(t) - x_0). \tag{11.2}$$

The intuitive explanation of this selection is that the more a participant wants to drive the partner away from the equilibrium, the stiffer the participant's arm should be in order to succeed. In other words, the robot with such a controller is compliant near the equilibrium while being stiffer when leaving it.

However, this implementation has a fundamental limitation for realizing full interactive handshakes as it lacks the ability to alter the reference trajectory and to switch to a passive mode. The robot can only playback motions as predefined, with the human input applied on top of it. This is clearly different from human-human handshaking, where the arms can provide compliance during interaction, while in the human mind different strategies can be selected about whether to adapt to the partner or not. On this account, an advanced more interactive handshaking controller was developed.

Advanced Handshake Model: Interactive Partner

The implementation of the interactive handshake partner builds on the idea that humans select between two different interaction strategies when performing handshakes with a partner: either they act passively by following and adapting their behavior as best as possible to the lead of the interaction partner; or they act actively by commanding the handshake trajectory without taking into account their partner's behavior. The recreation of handshakes is based on the idea of assigning the robot the respective opposite role/strategy by switching between pre-defined behavioral models that implement active and passive roles for the robot. Since the human interaction strategy changes over time, human interaction strategies need to be continuously estimated.

To implement this recreation strategy on a robotic system, a robot controller consisting of a control module, a planning module, and an adaptation module was realized, see Fig. 11.4. These three modules were already proposed in [46] for modeling a partner in a collaborating human-human dyad. While the planning unit decides on the desired trajectory, the control unit implements compliant behavior. The adaptation unit, finally, uses information about the actual estimated human interaction strategy to decide on different adaptation laws that alter the reference trajectory as well as the provided compliance of the robot. Thus, a new double-layered control scheme is proposed consisting of a low-level controller (LLC) that combines planning and control unit and a high-level controller (HLC), which is represented by the adaptation unit.

Human Behavior Model In order to recreate artificial handshakes following the aforementioned recreation strategy, information about the selected human interaction strategy is needed. Since these strategies are not directly measurable, they have

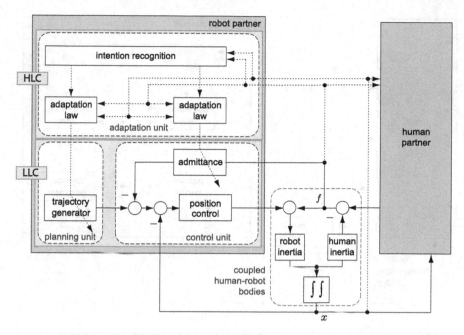

Fig. 11.4 Recreation of human-human handshakes using an advanced, interactive handshaking controller consisting of a low-level controller (LLC) that combines planning and control unit and a high-level controller (HLC) represented by the adaptation unit (first presented in [42])

to be estimated from observable variables of interaction, in our case haptic (force and motion) data. One of the challenges in this estimation is that force and motion signals contain information about both the human and the robot, hence any phenomenon observed in either signal can be the consequence of either the robot or the human. For this reason, motion or force signals alone cannot be adopted to estimate the currently selected human interaction strategy, but instead, need to be further processed to remove the influence of the robot. In order to overcome this problem, we introduced a human behavior model and estimated parameters of this model out of observable haptic interaction data. The estimated parameters are finally used as input for the human intention estimator described below.

While for the replay implementation, the human was assumed to be passively following the robot, which means that the human was modeled by a passive impedance represented by mass, damping, and stiffness of the human arm without further excitation signals to the system, for an active human the desired trajectory of the human becomes an additional input to the coupled human/robot system. Consequently, the old human model no longer applies and thus, a new human behavior model that implements the three units proposed in [46] is assumed:

1. The human is modeled by a position-controlled arm with a trajectory planner and an adaptation module that adjusts the compliance of the arm as well as the reference trajectory according to the actual estimated interaction strategy and the currently observed haptic data.

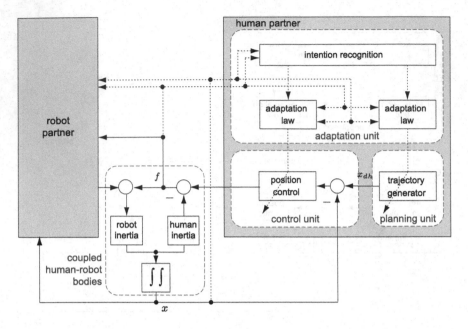

Fig. 11.5 Human behavior model. x_{dh} is the desired trajectory generated by the human, x is the actual human/robot position, and f is the actual interaction force. The parameters of human arm stiffness and damping represented by the position controller are variable in time (first presented in [42])

2. The human behaves as a collaborative partner. In other words, the human planner adapts the reference trajectory based on the actual interaction status. It does not matter whether the decision of the human is to follow or to change the current trajectory, the decision is made based on actual measurements.

   An illustration of the human behavior model is given in Fig. 11.5. This model allows the human to change the arm impedance represented by the position controller as well as the desired trajectory with respect to time. The assumption that the human will behave reasonably is a natural approach, since it is expected that a human plans the behavior of the next step based on the information gathered about the current step. Further information needed to generate the trajectory is integrated inside the planner block. This makes the current position and force signals the only inputs to the human model, hence the input and output signals of the human block are all known to the robot.

   The model in Fig. 11.5 is expected to be time-varying and non-linear. However, in practice the robot needs an easy-identifiable model for the interactive controller to estimate human intentions at each time instance. Hence the following approach was taken to linearize the human behavior model:

1. A linear differential equation was used to represent the relationship between position input and force output signals.

2. The differential equations were limited to the second order, as shown in (11.3), where f and x are the position input and force output signals, h_2, h_1, and h_0 are the three parameters of the differential equation, each denoted as a *human behavior parameter* (HBP), representing the current human behavior that determines the force output based on the current position input. Since the adaptation unit is part of the human behavior model the HBPs are time-varying.

$$f = h_2(t)\ddot{x} + h_1(t)\dot{x} + h_0(t)x. \tag{11.3}$$

The HBP set (h_2, h_1, h_0) is similar to the impedance parameter set of a passive human, in the sense that they are both relationships between force and position signals. However, HBPs also take into account the influence of the human planning and adaptation unit, and thus, are not necessarily equivalent to impedance parameters.

While details about the estimation of HBPs out of haptic data can be found in [31], the benefit of using HPB parameters as input for the human intention estimator instead of force and motion signals should be stressed. Given direct estimates of human behavior parameters from haptic data and knowing about the currently programmed robot behavior, the robot influence and thus, the bias in these parameters can be removed. Further, distinguishing between multiple human behavior parameters allows to disentangle ambiguities between a passive and active human, as a high value of h_0 indicates that the human is actively resisting the robot motion and thus, following an active strategy, while a high value of h_2 reflects the fact that the human is acting mainly as inertia and thus, trying to follow a passive strategy.

Human Intention Estimation   As the human intention is not directly measurable, a special estimator is needed. In our case we adopt pattern recognition algorithms which process the aforementioned HBPs and determine probabilities for the active or passive interaction strategy. A detailed scheme of the implemented human intention estimator, respectively interaction strategy estimator is shown in Fig. 11.6. Using measured force and motion data, an online parameter estimator identifies human behavior parameters, abstracts them into symbols and feeds them into a discrete Hidden-Markov-Model(HMM)-based intention recognition module which outputs an estimate of the current human intention, respectively interaction strategy. Two HMMs are defined for the estimator reflecting the two opposite roles *active* and *passive*. Active indicates that the human is trying to lead the handshake, while passive means the human likes to follow the lead of the robot. More details on the symbol abstraction and HMM strategy estimation can be found in [31, 42].

Trajectory and Impedance Adaptation   Depending on the estimated human intention the robot is programmed to take opposite roles by adapting the reference trajectories and by altering the provided compliance according to available adaptation laws.

The following strategy was used to implement active and passive robot behavior. When the human is in passive state, the robot tries to take the lead. Impedance parameters in the admittance filter are set to high values, and no modification is made to the current reference trajectory and the robot goes on as planned. When

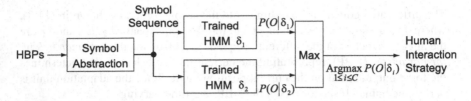

Fig. 11.6 Estimation of human interaction strategies using HMM-based estimator: HMM δ_1 stands for the active, and HMM δ_2 for the passive state. (Reprint with permission of [42])

the human is active, the robot tries to follow the lead. The impedance parameters are set to low values, and the commanded trajectory is modified such that the robot synchronizes to human shaking movements, which requires a continuous adaptation of the reference trajectory. In order to achieve such a behavior, different trajectory adaptation laws were tested. Our first implementation presented in [31] divided each handshake cycle into four segments and updated the planned trajectory at the beginning of each segment only. Thus, no matter how much effort the human applied to change the current trajectory during the segment, the effect only appeared when the next segment started. This delay of a quarter of a cycle left an irresponsive impression to the participants. In order to fix this latency issue, in [42] we proposed an advanced trajectory adaptation law, that takes information of every sample into account and adapts amplitude and frequency of the generated shaking motion based on currently observed position and force measurements in order to synchronize to the human.

Experiments showed that the roles active and passive are not fixed, but that partners switch between them several times. Using the aforementioned approach for human interaction strategy estimation, the number of detected switching events between the states active and passive can be easily modified by changing the thresholds of the discrete HMM. In doing so, an interaction partner with a controllable degree of dominance can be realized.

11.3.4 Summary

In this subsection a virtual handshake partner was designed and implemented following the *record-reply-recreate* workflow. Based on human-human handshake recordings, a basic controller was implemented that replays pre-recorded handshake trajectories on a robot. Since handshaking represents an interactive task, where partners mutually influence each other and adapt to each other, this handshake partner only resulted into very unnatural handshake behavior. To overcome this problem, an advanced, interactive handshake partner was developed that estimates human intentions online and adapts its planned trajectory and compliant behavior to the partner. Although the specific handshake model implementing the two states active and passive represents a very first, simple model of a handshake, it can be extended in many ways: more detailed states can be included and new adaptation laws can be implemented by taking into account recent behavioral and neuroscientific findings, see e.g. Chap. 6. In doing so, not only more realistic, but also more personalized, gender- or country-specific handshakes can be realized.

Finally, even though the handshake task is rather specific, the approach of modeling a handshake has huge potential for generalization. For instance, defining human intention states to represent human strategies of carrying out a task, using pattern recognition algorithms to estimate human intention states based on observable behavioral data, as well as online adaptation of robot strategies and parameters in order to change the quality of interaction are general research questions that generalize to applications like rehabilitation, walking assistants, robots that help in handling and transporting heavy and large objects as well as sensorimotor skill training.

11.4 Application Scenario: Dancing

In [47] we studied dancing as haptic interaction scenario. Like handshaking, dancing requires a physical coupling between partners, but differs from handshaking as (i) dominance is by definition distributed unequally between partners, (ii) the basic form of dancing steps is predefined, and (iii) dancing figures represent cyclical movements.

Several studies concerning the analysis of human behavior while dancing are known from literature. They can be split into two classes: the first class investigates human motion while performing a dance without haptic interaction, e.g. Japanese folk dance [48] or modern dance [49], and the second class investigates social dance with haptic interaction as analyzed in [50] where the haptic coordination between dancers with PHANToMs was examined and [51], where a female dancing robot was presented that follows the male, tries to estimate the next dancing step [20, 52] and is able to adapt the step size to the partner [53]. Our aim was to implement a haptic enabled male dancer that imitates the behavior of a real human partner. In contrast to female dancers which require the ability to estimate intentions, male dancers need the ability to communicate intentions and to adapt to their partner's behavior.

In order to achieve this behavior, we followed again the *record-replay-recreate* workflow introduced above. First, we observed the physical interaction of two humans to *record* movements and forces occurring while dancing. The acquired data sets describe specific interactions which are unique for the recorded couples and trials. A robot, that simply *replays* data sets, can only perform these recorded motions. The fundamental characteristics of dancing, (i) arbitrary sequences of dancing steps and (ii) slight variations of the dancing steps depending on external influences, require a model of the recorded interactions to *recreate* dancing behavior.

The following two sections describe the record and the replay of dancing sequences, followed by our approach of recreating an interactive male robot dancer.

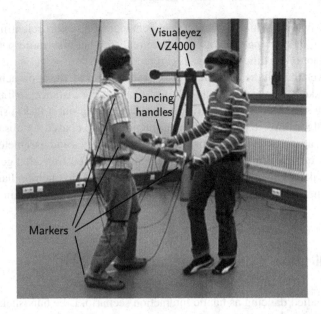

Fig. 11.7 Recording of dancing couple: Motion capture system and dancing handles with force–torque sensors

11.4.1 Recording

In the recording phase nine semi-professional dancing couples were recorded using a motion capture system Visualeyez VZ4000 system at a frame rate of 25 Hz, see Fig. 11.7. A number of 42 markers were placed on special body parts like head, shoulders, elbows, hands, hips, knees and feet.

The Discofox was chosen as dance as it allows to have only two interaction points. In order to measure the interaction force, special adapters were constructed which connect two handles over a 6 DOF force-torque sensor (100 Hz) and consequently allow dancing-like hand postures. The position and orientation of each handle was tracked with four motion capture markers placed around the sensor, see Fig. 11.8.

The model of the virtual male dancer was built based on the recorded positions and forces. The additional markers placed on the bodies of the dancers recorded their movements. This information was later used for the animation of the virtual dancing partner presented to the user through a head-mounted display.

In order to take the dynamic capabilities and constraints of the robot into account, the song "Tears in Heaven" with 79 beats per minute was selected to record three different dancing steps: basic step (forward/backward), side step, and windmill (rotation). To achieve synchronicity between motion capture and force-torque signals, as well as music, one central computer recorded the data and generated the beat for the MIDI player which played the music.

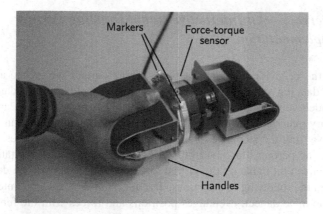

Fig. 11.8 Dancing handle with force-torque sensor and markers

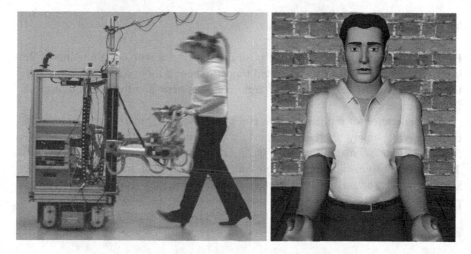

Fig. 11.9 Dancing with mobile platform (*left*), virtual avatar seen through head-mounted display (*right*)

11.4.2 Replay—A Leading Partner Implementation

In the replay phase the recorded motion data was replayed on a mobile robotic platform [54] equipped with two robotic arms [55], Fig. 11.9 (left). The mobile platform was position controlled and the robotic arms were programmed to follow the measured positions of left and right hand recorded during human-human interaction, while the head-mounted display presented an animated virtual avatar, see Fig. 11.9 (right). Adopting this approach led to a fully dominant male dancer that cannot adapt to the female partner.

11.4.3 Recreate—An Interactive Robot Dancing Partner Implementation

To recreate the behavior of an interactive dancing partner we had to address the planning of dancing steps and the adaptation of these steps to the dancing partner. The model we used to realize a dancing partner is based on [32] and consists of a trajectory generator for the synthesis of the dancing steps and an adaptation module for varying the step size based on measured interaction forces. Similar to the handshaking partner illustrated in Fig. 11.4, a planning and adaptation unit as well as a control unit make up the virtual male partner, which will be described in more detail in the following two sections. The admittance-type haptic interface used to replace the male dancer, exchanges positions and forces with the female, while the implemented position-based admittance controller implements compliance and thus, allows small deviations during interaction.

Trajectory Generator

Human movements like dancing are position trajectories with certain points of the body moving over time. The generated trajectories of a robot dancing partner have to allow transitions between the different dancing steps and the adaptation to the female partner.

The generation of trajectories is a common problem in motion imitation applications. While many approaches consider the generation of trajectories [38, 49, 56, 57], only a few methods take physical interaction into account [22, 43, 58, 59]. The non-linear dynamics approach presented by Okada et al. [60] allows to synthesize motions based on a given trajectory while reducing the amount of needed data points.

The desired dancing trajectory C is considered to be given as M position samples c measured over time:

$$C = \begin{bmatrix} c[1] & c[2] & \cdots & c[M] \end{bmatrix} \tag{11.4}$$

$$c[k] = \begin{bmatrix} c_1[k] & c_2[k] & \cdots & c_N[k] \end{bmatrix}^T \tag{11.5}$$

with

$$c[M+1] = c[1] \tag{11.6}$$

to match the requirement of a cyclic task.

The non-linear dynamical system of the form

$$x[k+1] = x[k] + f(x[k]) \tag{11.7}$$

can describe the trajectories using the system dynamics f. The discrete-time state vector $x[k]$ usually contains position and velocity. With a constant sampling time, the system stores the velocity of the trajectory implicitly.

Fig. 11.10 Recorded
trajectory with vector field

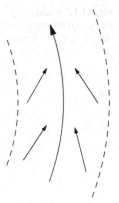

Fig. 11.11 Attracting vectors
around trajectory

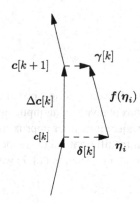

A vector field is constructed around the recorded trajectory as shown in
Fig. 11.10. To assure an attracting behavior the vectors are pointing towards the
trajectory.

Around each measured data point $c[k]$ of the trajectory, several points η_i are
defined with $\delta[k]$ and $\gamma[k]$ perpendicular to $\Delta c[k]$, see Fig. 11.11.

The system dynamics $f(x[k])$ can be determined by approximating the vector
field $f(\eta_i)$ using a polynomial expression with the degree l:

$$f(\eta_i) = \Phi \theta(x) \tag{11.8}$$

$$\theta(x) = \begin{bmatrix} x_1^l & x_1^{l-1} x_2 & x_1^{l-2} x_2^2 & \cdots & 1 \end{bmatrix}^T \tag{11.9}$$

The matrix Φ consists of constant parameters and is determined with the least square
method:

$$\Phi = F \Theta^{\#} \tag{11.10}$$

$$F = \begin{bmatrix} f(\eta_1) & f(\eta_2) & \cdots & f(\eta_n) \end{bmatrix} \tag{11.11}$$

$$\Theta = \begin{bmatrix} \theta(\eta_1) & \theta(\eta_2) & \cdots & \theta(\eta_n) \end{bmatrix} \tag{11.12}$$

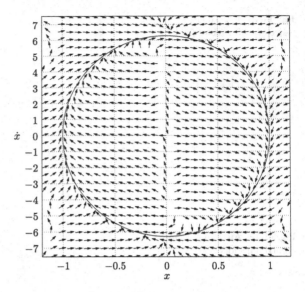

Fig. 11.12 Exemplary sinusoidal movement with attracting vector field

Okada et al. proposed an extension to influence the system dynamics with different levels of an input signal. Thus, the dynamical system can synthesize different trajectories after approximating the recorded trajectory which has been segmented into different states, respectively dancing steps in our application. The dynamical system shown in (11.7) was extended to

$$x[k+1] = x[k] + g(u[k], x[k]),　　　　　　　　　(11.13)$$

where the $u[k]$ denotes the input signal which is appended to C in (11.4). An exemplary sinusoidal movement and its vector field are shown in Fig. 11.12.

The dynamical system requires cyclic movements matching the time period which is not always fulfilled for all degrees of freedom of a 6 DOF application like dancing. The trajectory of the basic step for example as shown in Fig. 11.13 is cyclic in forward/backward direction, but the lateral movement and the height oscillate around a mean value with arbitrary patterns.

To overcome this problem, each DOF of the recorded trajectories was modulated on a sinusoidal movement. The radius of the pointer on the circle denotes the amplitude while the angle represents the timing of the dancing step, see Fig. 11.14.

The system dynamics can be scaled with a scalar to assure a continuous synchronization between dynamical system and music. The factor slightly accelerates or decelerates the dynamical system depending on the phase shift between the generated signal and the beat of the music.

The recorded and synthesized movements in forward/backward direction of the step sequence *forward/backward–side* are shown in Fig. 11.15.

Figures 11.16 and 11.17 show the corresponding phase plots and vector fields of the modulated signal for the *forward/backward* (outer curve) and *side* step (inner curve) respectively.

Fig. 11.13 Movements of
left male hand for recorded
dancing step
forward/backward:
Forward/backward direction
(*solid*), left/right movement
(*dashed*), and up/down
movement (*dash-dot*)

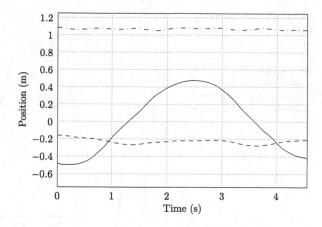

Fig. 11.14 Modulation of
movement on circle

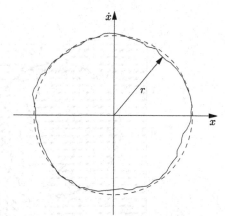

Trajectory Adaptation

The interaction between the dancing partners requires certain variations of the dancing steps even if the male dancing partner leads the dance. Based on this work, Sato
et al. [35] implemented a human-robot interaction algorithm that synchronizes and
adapts between two systems taking handshaking as example. Since they considered handshaking a task requiring mutual adaptation between two partners, they
implemented an adaptation procedure based on the method of online design of dynamics, which results in an implicit adaptation. For handshaking, such behavior is
desired as the partners are allowed to converge to an arbitrary common amplitude
and frequency. However in dancing, movements and especially their timing are subject to constraints: The frequency has to stay constant in order to keep the beat of
the music. While the adaptation approach realized by Sato et al., cannot deal with
such constraints, our approach implements an explicit adaptation that builds on the
already mentioned non-linear dynamics approach of the trajectory generator. Exemplarily chosen trajectories of different dancing steps were synthesized and represent
a totally dominant male partner. The generated trajectories were then scaled by the

Fig. 11.15 Movement in
forward/backward direction
of the step sequence
forward/backward-side;
recorded (*solid*) and
synthesized (*dashed*)
trajectory; *gray area* marks
side step

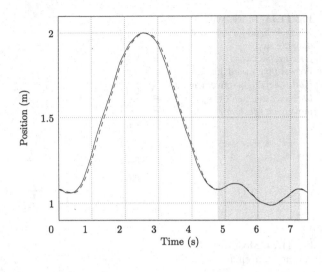

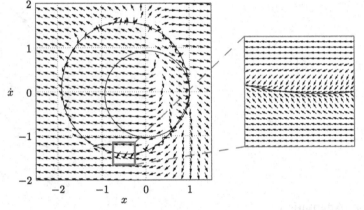

Fig. 11.16 Phase plot with vector field around *forward/backward* step (*outer curve*)

adaptation module according to the currently measured interaction force between
the participants. The force between female and haptic interface served as measure
for the preferred size of the dancing steps. A pulling (pushing) force while the female is moving forward (backward) indicated that the dancing steps are too large
(small). Thus, the step size adaptation module had to adapt the size of the dancing
step to reduce the interaction force between the female and the haptic interface.

The output $x[k]$ of the dynamical system (11.13) is scaled

$$y[k] = a[k]x[k] + d[k], \qquad (11.14)$$

with the factor $a[k]$, while $y[k]$ denotes the output of the adaptation and $d[k]$ cor-
rects occurring offset to avoid jumps and drift effects.

The forward/backward movement of the forward/backward step are chosen to
illustrate the adaptation of the robot to the female. The mean force between female

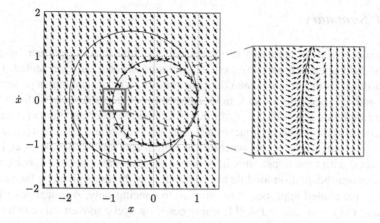

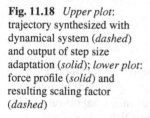

Fig. 11.17 Phase plot with vector field around *side* step (*inner curve*)

Fig. 11.18 *Upper plot*: trajectory synthesized with dynamical system (*dashed*) and output of step size adaptation (*solid*); *lower plot*: force profile (*solid*) and resulting scaling factor (*dashed*)

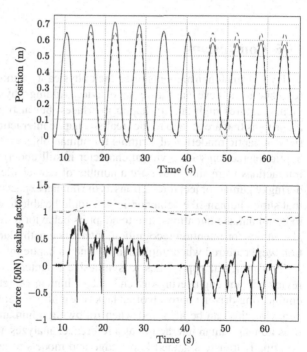

and robot multiplied by the sign of the robot's velocity is shown in the lower plot of Fig. 11.18. Due to different signs of velocity and force, the interaction force profile is lower than zero during the first 60 seconds which leads to a reduced step size. The subsequent interaction force profile is positive resulting in larger dancing steps. The asymmetric offset of the adapted position results from $d[k]$.

11.4.4 Summary

In this subsection an approach to synthesize a virtual dancing partner based on a vector field approach and an explicit adaptation strategy was presented. Based on recordings from human-human dancing couples, typical dancing steps were extracted and encoded into a vector field. Adaptation of step sizes was realized by an explicit adaptation law, which took the actual interaction force and human intentions into account. While other approaches adopting the vector field approach realized an implicit adaptation strategy to synchronize human and robot movements and thus, cannot cope with time constraints imposed by the dancing application, our explicit and continuous adaptation module allowed to follow the rules given by the task.

The implemented approach is not limited to dancing only. Applications like rehabilitation or robot-assisted skill learning based on cyclic movements can be modelled in a similar way. For future versions of the system, a time-variable admittance of the compliant controller or a variable adaptation rule for the trajectory generator are the most promising extensions.

11.5 Conclusions

Today's virtual environments often lack physicality although it is considered a prerequisite to achieve natural user behavior. Adding physicality to virtual environments requires high-quality haptic interfaces, but also advanced haptic rendering algorithms that are able to render realistic haptic interactions. While in the past a variety of haptic rendering algorithms for human-object interactions were developed, haptic interaction with a virtual character is still underinvestigated. Such physical interactions were shown to pose a number of new challenges compared to the rendering of human-object interactions as the human expects to interact with a character that shows human-like behavior, i.e. it should be able to estimate human intentions, to communicate intentions, and to adapt its behavior to its partner. On this account, algorithms for intention recognition, interactive path planning, control, and adaptation were required when implementing such a haptic interaction partner.

Two principal approaches to synthesize an interactive, virtual character were reviewed: an engineering-driven and a human-centered approach. While the engineering-driven approach often lacks the ability to realize human-like interaction behavior, this can be achieved when following a human-centered approach which uses human-human interaction as a reference, analyzes and models this interaction, and finally transfers gained knowledge and models to human-robot interaction. In the presented work we followed this latter approach and adopted a three-step workflow implementing a record, replay, and recreate phase to realize a virtual, haptic interaction partner. The validity of the introduced approach was demonstrated in two prototypical application scenarios, handshaking and dancing.

Acknowledgements   This work was partly supported by the ImmerSence project within the 6th Framework Programme of the European Union, FET—Presence Initiative, contract number IST-2006-027141, see also www.immersence.info.

References

1. Burdea, G.: Force and Touch Feedback for Virtual Reality. Wiley, New York (1996)
2. Martin, J., Savall, J.: Mechanisms for haptic torque feedback. In: Proc. of the First Joint Eurohaptics Conference and Symposium on Haptic Interfaces for Virtual Environment and Teleoperator Systems, pp. 611–614 (2005)
3. Zilles, C.B., Salisbury, J.K.: A constraint-based god-object method for haptic display. In: Proc. of the IEEE/RSJ International Conference on Intelligent Robots and Systems (1995)
4. Basdogan, C., Srinivasan, M.A.: Haptic rendering in virtual environments. In: Stanney, K.M. (ed.) Handbook of Virtual Environments: Design, Implementation, and Applications, pp. 117–134. Lawrence Erlbaum Associates, Mahwah (2002)
5. Thompson, T.V. II, Johnson, D.E., Cohen, E.: Direct haptic rendering of sculptured models. In: Proceedings Symposium on Interactive 3D Graphics (1997)
6. Kim, L., Kyrikou, A., Sukhatme, G.S., Desbrun, M.: An implicit-based haptic rendering technique. In: Proc. of the IEEE/RSJ International Conference on Intelligent Robots and Systems (2002)
7. Okamura, A.M., Kuchenbecker, K.J., Mahvash, M.: Measurement-based modeling for haptic rendering. In: Haptic Rendering: Foundations, Algorithms, and Applications. AK Peters Series (2008)
8. Basdogan, C., Ho, C.H., Srinivasan, M.A.: A ray-based haptic rendering technique for displaying shape and texture of 3D objects in virtual environments. In: Dynamic Systems and Control Division (1997)
9. Otaduy, M.A., Lin, M.C.: Rendering of textured objects. In: Haptic Rendering: Foundations, Algorithms, and Applications. AK Peters Series (2008)
10. Lin, M., Otaduy, M. (eds.): Haptic Rendering, Foundations, Algorithms, and Applications. AK Peters Series (2002)
11. Hirata, Y., Kosuge, K.: Distributed robot helpers handling a single object in cooperation with humans. In: Proc. of the IEEE International Conference on Robotics and Automation, pp. 458–463 (2000)
12. Rahman, M.M., Ikeura, R., Mizutani, K.: Control characteristics of two humans in cooperative task and its application to robot control. In: 26th Annual Conference of the IEEE Industrial Electronics Society, pp. 1773–1778 (2000)
13. Tsumigawa, T., Yokogawa, R., Hara, K.: Variable impedance control with regard to working process for man-machine cooperation-work system. In: Proc. of the IEEE/RSJ International Conference on Intelligent Robots and Systems (2001)
14. Tsumigawa, T., Yokogawa, R., Hara, K.: Variable impedance control with virtual stiffness for human-robot cooperative peg-in-hole task. In: IEEE/RSJ International Conference on Intelligent Robots and Systems, pp. 1075–1081 (2002)
15. Tsumigawa, T., Yokogawa, R., Hara, K.: Variable impedance control based on estimation of human arm stiffness for human-robot cooperative calligraphic task. In: IEEE International Conference on Robotics and Automation, pp. 644–650 (2002)
16. Ikeura, R., Moriguchi, T., Mizutani, K.: Optimal variable impedance controller for a robot and its application to lifting an object with a human. In: IEEE International Workshop on Robot and Human Interactive Communication, pp. 500–505 (2002)
17. Duchain, V., Gosselin, C.M.: General model of human-robot cooperation using a novel velocity based variable impedance control. In: Worldhaptics, Tsukuba, Japan, pp. 446–451 (2007)
18. Arai, H., Takubo, T., Hayashibara, Y., Tanie, K.: Human-robot cooperative manipulation using a virtual nonholonomic constraint. In: IEEE Proc. of the International Conference on Robotics and Automation vol. 4, pp. 4063–4069 (2000). doi:10.1109/ROBOT.2000.845365
19. Maeda, Y., Hara, T., Arai, T.: Human-robot cooperative manipulation with motion estimation. In: Proc. of the IEEE/RSJ International Conference on Intelligent Robots and Systems, Maui, Hawaii, pp. 2240–2245 (2001)

20. Takeda, T., Hirata, Y., Kosuge, K.: Dance step estimation method based on HMM for dance partner robot. IEEE Trans. Ind. Electron. **54**(2), 699–706 (2007)
21. Corteville, B., Aertbelien, E., Bruyninckx, H., de Schutter, J., van Brussel, H.: Human-inspired robot assistant for fast point-to-point movements. In: IEEE International Conf. on Robotics and Automation, Roma, Italy, pp. 3639–3644 (2007)
22. Evrard, P., Gribovskaya, E., Calinon, S., Billard, A., Kheddar, A.: Teaching physical collaborative tasks: Object-lifting case study with a humanoid. In: 9th IEEE-RAS International Conference on Humanoid Robots, pp. 399–404 (2009)
23. Wojtara, T., Uchihara, M.M.H., Shimoda, S., Sakai, S., Fujimotor, H., Kimura, H.: Human-robot collaboration in precise positioning of a three-dimensional object. Automatica **45**, 333–342 (2009)
24. Rahman, M., Ikeura, R., Mizutani, K.: Cooperation characteristics of two humans in moving an object. Mach. Intell. Robot. Control **4**, 43–48 (2002)
25. Reed, K.B., Peshkin, M., Hartmann, M.J., Patton, J., Vishton, P.M., Grabowecky, M.: Haptic cooperation between people, and between people and machines. In: Proceedings of the 2006 IEEE/RSJ Conference on Intelligent Robots and Systems, Beijing, China (2006)
26. Reed, K.B., Peshkin, M.A.: Physical collaboration of human-human and human-robot teams. IEEE Trans. Haptics **1**, 108–120 (2008)
27. Miossec, S., Kheddar, A.: Human motion in cooperative tasks: Moving object case study. In: Proceedings of the 2008 IEEE International Conference on Robotics and Biomimetics (2008)
28. Feth, D., Groten, R., Peer, A., Hirche, S., Buss, M.: Performance related energy exchange in haptic human-human interaction in a shared virtual object manipulation task. In: Third Joint EuroHaptics Conference and Symposium on Haptic Interfaces for Virtual Environment and Teleoperator Systems (2009)
29. Groten, R., Feth, D., Goshy, H., Peer, A., Kenny, D.A., Buss, M.: Experimental analysis of dominance in haptic collaboration. In: The 18th International Symposium on Robot and Human Interactive Communication (2009)
30. Groten, R., Feth, F., Klatzky, R., Peer, A., Buss, M.: Efficiency analysis in a collaborative task with reciprocal haptic feedback. In: The 2009 IEEE/RSJ International Conference on Intelligent Robots and Systems (2009)
31. Wang, Z., Peer, A., Buss, M.: An HMM approach to realistic haptic human-robot interaction. In: Worldhaptics (2009)
32. Hölldampf, J., Peer, A., Buss, M.: Virtual partner for a haptic interaction task. In: Ritter, H., Sagerer, G., Dillmann, R., Buss, M. (eds.) Human Centered Robot Systems. Cognitive Systems Monographs, pp. 183–191. Springer, Berlin (2009)
33. Kunii, Y., Hashimoto, H.: Tele-handshake using handshake device. In: Proceedings of the 21st IEEE International Conference on Industrial Electronics, Control, and Instrumentation, USA, vol. 1, pp. 179–182 (1995)
34. Pollard, N.S., Zordan, V.B.: Physically based grasping control from example. In: Proceedings of the 2005 ACM SIGGRAPH/Eurographics symposium on Computer Animation, USA, pp. 311–318 (2005)
35. Sato, T., Hashimoto, M., Tsukahara, M.: Synchronization based control using online design of dynamics and its application to human-robot interaction. In: Proceedings of the IEEE International Conference on Robotics and Biomimetics, pp. 652–657 (2007)
36. Yamato, Y., Jindai, M., Watanabe, T.: Development of a shake-motion leading model for human-robot handshaking. In: Proceedings of the SICE Annual Conference 2008, Japan, pp. 502–507 (2008)
37. Atkeson, C.G., Schaal, S.: Robot learning from demonstration. In: Proceedings of the 14th International Conference on Machine Learning (1997)
38. Ijspeert, A.J., Nakanishi, J., Schaal, S.: Trajectory formation for imitation with nonlinear dynamical systems. In: Proc. IEEE/RSJ Int Intelligent Robots and Systems Conf., vol. 2, pp. 752–757 (2001). doi:10.1109/IROS.2001.976259
39. Kormushev, P., Calinon, S., Caldwell, D.G.: Imitation learning of positional and force skills demonstrated via kinesthetic teaching and haptic input. Adv. Robot. **25**(5), 581–603 (2010)

40. Schmidts, A.M., Lee, D., Peer, A.: Imitation learning of human grasping skills from motion and force data. In: IEEE/RSJ International Conference on Intelligent Robots and Systems (2011)
41. Solis, J., Marcheschi, S., Frisoli, A., Avizzano, C., Bergamasco, M.: Reactive robot system using a haptic interface: and active interaction to transfer skills from the robot to unskilled persons. Adv. Robot. 21(3–4), 267–291 (2007)
42. Wang, Z., Giannopoulos, E., Slater, M., Peer, A., Buss, M.: Handshake: Realistic human-robot interaction in haptic enhanced virtual reality. Presence (2011, accepted)
43. Calinon, S., Evrard, P., Gribovskaya, E., Billard, A., Kheddar, A.: Learning collaborative manipulation tasks by demonstration using a haptic interface. In: 14th International Conference on Advanced Robotics (ICAR), Munich, June 2009
44. Wang, W., Hölldampf, J., Buss, M.: Design and performance of a haptic data acquisition glove. In: Proceedings of the 10th Annual International Workshop on Presence, Spain, pp. 349–357 (2007)
45. Wang, Z., Yuan, J., Buss, M.: Modelling of human haptic skill: a framework and preliminary results. In: Proceedings of the 17th IFAC World Congress, Korea, pp. 14761–14766 (2008)
46. Groten, R.: Haptic human-robot collaboration: How to learn from human dyads. PhD thesis, Technische Universität München (2011)
47. Hölldampf, J., Peer, A., Buss, M.: Synthesis of an interactive haptic dancing partner. In: Proc. IEEE RO-MAN, pp. 527–532 (2010). doi:10.1109/ROMAN.2010.5598616
48. Nakazawa, A., Nakaoka, S., Ikeuchi, K., Yokoi, K.: Imitating human dance motions through motion structure analysis. In: Proc. IEEE/RSJ Int Intelligent Robots and Systems Conf., vol. 3, pp. 2539–2544 (2002). doi:10.1109/IRDS.2002.1041652
49. Pullen, K., Bregler, C.: Motion capture assisted animation: texturing and synthesis. In: SIGGRAPH'02: Proceedings of the 29th Annual Conference on Computer Graphics and Interactive Techniques, New York, NY, USA, pp. 501–508. ACM, New York (2002). doi:10.1145/566570.566608
50. Gentry, S., Murray-Smith, R.: Haptic dancing: human performance at haptic decoding with a vocabulary. In: Proc. IEEE Int Systems, Man and Cybernetics Conf., vol. 4, pp. 3432–3437 (2003). doi:10.1109/ICSMC.2003.1244420
51. Kosuge, K., Hayashi, T., Hirata, Y., Tobiyama, R.: Dance partner robot -Ms DanceR-. In: Proc. IEEE/RSJ Int. Conf. Intelligent Robots and Systems (IROS 2003), vol. 4, pp. 3459–3464 (2003). doi:10.1109/IROS.2003.1249691
52. Hirata, Y., Hayashi, T., Takeda, T., Kosuge, K., Wang, Z.-d.: Step estimation method for dance partner robot "MS DanceR" using neural network. In: Proc. 2005 IEEE Int. Conf. Robotics and Biomimetics (ROBIO), pp. 523–528 (2005). doi:10.1109/ROBIO.2005.246322
53. Takeda, T., Hirata, Y., Kosuge, K.: Dance partner robot cooperative motion generation with adjustable length of dance step stride based on physical interaction. In: Proc. IEEE/RSJ Int. Conf. Intelligent Robots and Systems IROS 2007, pp. 3258–3263 (2007). doi:10.1109/IROS.2007.4399270
54. Unterhinninghofen, U., Schauss, T., Buss, M.: Control of a mobile haptic interface. In: Proc. IEEE Int. Conf. Robotics and Automation ICRA 2008, pp. 2085–2090 (2008). doi:10.1109/ROBOT.2008.4543514
55. Peer, A., Buss, M.: A new admittance type haptic interface for bimanual manipulations. IEEE/ASME Trans. Mechatron. 13(4), 416–428 (2008)
56. Degallier, S., Santos, C.P., Righetti, L., Ijspeert, A.: Movement generation using dynamical systems: a humanoid robot performing a drumming task. In: Proc. 6th IEEE-RAS Int Humanoid Robots Conf., pp. 512–517 (2006). doi:10.1109/ICHR.2006.321321
57. Gribovskaya, E., Billard, A.: Learning nonlinear multi-variate motion dynamics for real-time position and orientation control of robotic manipulators. In: Proc. 9th IEEE-RAS Int. Conf. Humanoid Robots Humanoids 2009, pp. 472–477 (2009). doi:10.1109/ICHR.2009.5379536
58. Reed, K.B., Patton, J., Peshkin, M.: Replicating human-human physical interaction. In: Proc. IEEE Int Robotics and Automation Conf., pp. 3615–3620 (2007). doi:10.1109/ROBOT.2007.364032

59. Lee, D., Ott, C., Nakamura, Y.: Mimetic communication with impedance control for physical human-robot interaction. In: Proc. IEEE Int. Conf. Robotics and Automation ICRA '09, pp. 1535–1542 (2009). doi:2009.5152857
60. Okada, M., Tatani, K., Nakamura, Y.: Polynomial design of the nonlinear dynamics for the brain-like information processing of whole body motion. In: Proc. IEEE Int. Conf. Robotics and Automation ICRA'02, vol. 2, pp. 1410–1415 (2002). doi:10.1109/ROBOT.2002.1014741

Chapter 12
FMRI Compatible Sensing Glove for Hand Gesture Monitoring

Nicola Vanello, Valentina Hartwig, Enzo Pasquale Scilingo, Daniela Bonino, Emiliano Ricciardi, Alessandro Tognetti, Pietro Pietrini, Danilo De Rossi, Luigi Landini, and Antonio Bicchi

Abstract Here we describe and validate a fabric sensing glove for hand finger movement monitoring. After a quick calibration procedure, and by suitably processing of the outputs of the glove, it is possible to estimate hand joint angles in real time. Moreover, we tested the fMRI compatibility of the glove and ran a pilot fMRI experiment on the neural correlates of handshaking during human-to-human and human-to-robot interactions. Here we describe how the glove can be used to monitor correct task execution and to improve modeling of the expected hemodynamic responses during fMRI experimental paradigms.

12.1 Introduction

Hand and finger dynamics monitoring can be exploited for display and control purposes, to reproduce a realistic behavior in robotic and virtual reality systems, and to design behavioral and functional studies on cognitive and motor control of hand-executed tasks. Even if several systems are available to monitor hand movements, one of the main limitations of the currently available technologies is represented by the mechanical constraints, that may affect or hinder task execution. Our group has proposed a novel sensing glove realized with strain sensing fabrics [1]. The glove can be worn without any discomfort for the subject and does not interfere with hand

N. Vanello (✉) · L. Landini
Department of Information Engineering, University of Pisa, via G. Caruso, 16, 56122 Pisa, Italy
e-mail: nicola.vanello@iet.unipi.it

V. Hartwig
Institute of Clinical Physiology, National Research Council, via Moruzzi, 1, 56124 Pisa, Italy

V. Hartwig · E.P. Scilingo · A. Tognetti · D. De Rossi · A. Bicchi
Interdepartmental Research Center "E. Piaggio", University of Pisa, via Diotisalvi, 2, 56126 Pisa, Italy

D. Bonino · E. Ricciardi · P. Pietrini
Laboratory of Clinical Biochemistry and Molecular Biology, University of Pisa Medical School, Via Roma, 67, 56126 Pisa, Italy

A. Peer, C.D. Giachritsis (eds.), *Immersive Multimodal Interactive Presence*,
Springer Series on Touch and Haptic Systems,
DOI 10.1007/978-1-4471-2754-3_12, © Springer-Verlag London Limited 2012

movements. This property is achieved by employing sensors and connections made by elastomeric material that allow to maintain the elastic properties of the textile substrate in which they are integrated. Different sensor data processing strategies and calibration procedures can be adopted depending upon the specific application of the system. By using this glove we can monitor hand posture and gestures, by training a classifier with the set of sensor outputs obtained with different fingers positions, while by using a calibrating device as a reference for actual angles, it is possible to provide finger joint angles estimates in real time. Moreover, the glove is made by non ferromagnetic materials, and employs low intensity currents, thus results in a good compatibility with functional Magnetic Resonance Imaging (fMRI) studies. This application can be useful to monitoring haptic and motor tasks during brain functional studies, improving the identification and characterization of brain areas subserving the execution of specific hand-mediated movements. In fact fMRI data analysis relies on the definition of the Blood Oxygenation Level Dependent (BOLD) response, that can be modeled using the exact timing as well as finger dynamics, i.e. joint angles range or information about hand shape. During the last ten years, several devices have been developed for this purpose [2–7]. These devices have to satisfy safety and compatibility criteria. Recently, more accurate testing procedures have been adopted to verify whether these devices do affect image quality. In particular our group developed a novel, versatile, and automatic method that offers a statistically based approach to assess image quality for fMRI studies with mechatronic devices [8].

Furthermore, we will describe a preliminary study on the neural correlates of a multimodal social interaction in handshaking while using our sensing glove [9]. The proposed monitoring system will be used to investigate the role of haptic and vision during the execution of handshaking for both characterizing human-to-human interaction and highlight the differences between human-to-human and human-to-robot interaction (HRI). This study aims at further clarifying integration processes of multimodal information in humans and may provide useful information for the design of social robots in real life as well as in virtual reality applications.

12.2 Wearable Sensing Glove

12.2.1 System Description

The sensors of the glove are made of a conductive elastomer (CE) material, provided by Wacker LTD, which show piezoresistive properties, i.e. it changes its resistance when a deformation is applied. The sensors are printed on a Lycra/cotton fabric without affecting its elastic properties. In Fig. 12.1 the thicker line represents the sensors connected in series.

A constant current lower than 50 µA is injected through this sensors path. When a subject wearing the glove flexes a finger, thus stretching one or more sensors, a voltage change occurs between the sensors terminals. Thinner lines represent connections from sensors to the instrumentation amplifier, thus allowing to limit the

Fig. 12.1 The wearable sensing glove. *Black lines* are made by conductive elastomer

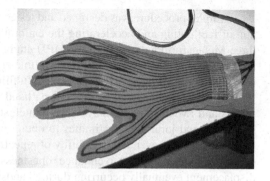

Fig. 12.2 The CE sensors path (*thicker line*) and connections are shown. This topology allows to realize 20 sensors

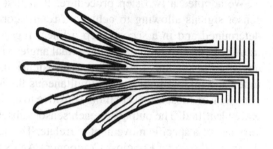

value of the current flowing through them ($\sim$ nA). This small current level, orders of magnitude smaller than the sensors current, makes it possible to ignore the differences in voltage drop due to connections resistance changes. The system is realized with 20 sensors connected in series and with the topology shown in Fig. 12.2.

A copper shielded cable is used to connect through a waveguide, the sensing glove to the acquisition electronics and to the current generator, which are located far away the sensing glove in the console room.

Calibration Procedures

The sensor data can be processed using different approaches depending upon the specific needs. For the purpose of hand posture recognition the set of voltages through the sensors, corresponding to different hand positions, can be stored. A dataset obtained with 32 different hand positions, was used to train a classifier with an Euclidean distance minimization algorithm and a complete recognition success. The removal of the glove and its re-worn usually reduces the performance of the classifier (approximately 5% less), given possible sensor displacements with respect to positions obtained during the training phase. In [1] a feedforward network was trained to estimated joint angles using the Cyberglove© (produced by Immersion LTD), as a validation device. The Cyberglove and our system were worn at the same time both during the training and the test phase. A maximum error smaller than 10% was found.

A simpler procedure was designed and tested for joint angles estimation, with the aim of facilitating and accelerating the calibration phase. The goal was to achieve three Metacarpo Phalangeal Joint (MPJ) angle estimates in real time, with an error smaller than 10%, with a calibration time smaller than one minute and without using any external reference devices. The fulfillment of these general requirements was suggested by the need of monitoring hand dynamics during handshake experiments and fMRI. In particular we were interested in exact movement timings (i.e. start/stop) and joint angle estimates to obtain hand aperture (small/wide aperture) information. Moreover, the possibility of repeating calibration procedure during the experiment was important to enhance robustness of the results with respect to sensor displacement eventually occurring during handshake.

We adopted a two step procedure: in a first step a suitable transformation of sensor signals allowing to achieve a robust correlate of specific movements was determined and in a second step a linear regression model was used to map the transformed sensors signal to the actual angles. The goal was to estimate three MPJ angles corresponding to thumb flexion/extension, simultaneous flexion/extension of index and middle fingers and simultaneous flexion/extension of ring and little fingers. Sensor subsets that are more sensitive to each of the three classes of movements were identified. The outputs of each sensor subset were then combined to maximize variance of a specific movement correlate. The sensor weights for each subset were determined using a Principal Component Analysis of the sensor outputs, retaining only the first component. A linear regression was applied to each sensor subset using the principal component score as independent variable and the actual angles as dependent variable. This second step was realized using known angle positions, 0 degrees or flat hand position, with fingers close to each other, and two 90 degree positions, one with the thumb touching the palm of the hand with the other fingers in flat position, and one with the thumb in flat position and the other fingers clenched in a fist. To test this procedure we performed tests using electrogoniometers to measure desired angles. Figures 12.3 and 12.4 show the results related to thumb and index/middle fingers. Using this approach a maximum measurement error of 10 degrees was observed and better performances were obtained during the finger closing phase, i.e. in sensor stretching, with respect to the finger opening phase.

12.3 fMRI Compatibility

12.3.1 A Statistically Based Approach

Compatibility and safety criteria must be satisfied in designing of every new device able to convey different sensory stimuli to the subject inside the MR scanner environment during fMRI studies. Hence, interaction between MR primary components and mechatronic devices has to be considered [10–12]. A fMRI compatibility test was applied to our glove, which employs mean differences of image quality indexes between selected experimental conditions (i.e., we compared MRI sequences

Fig. 12.3 Angle test measurement. Electrogoniometer (*dot-dashed line*) and our system estimated MPJ angle for index/middle fingers are shown with respect to time (s)

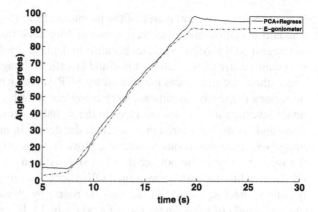

Fig. 12.4 Angle test measurement. Electrogoniometer (*dot-dashed line*) and our system estimated MPJ angle for thumb finger are shown with respect to time (s). Reproduced with permission from [9] © 2010 IEEE

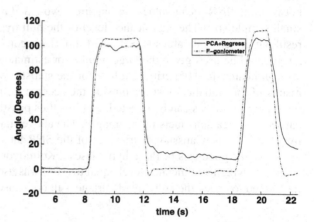

recorded while the device is inside the scan environment and sequences in a control condition without the device or baseline condition). The BOLD sensitive T2*-weighted images are GE-EPI [echo time/repetition time (TE/TR) = 40/3000 ms, flip angle (FA) = 90°, bandwidth (BW) = 62.5 kHz, field of view (FOV) = 24 cm, number of slices = 25, 5-mm-thick axial slices, 1.5 T GE Scanner Excite HD] acquired using a spherical phantom of $CuSO4$ solution. As first image quality index the Signal to Noise Ration (SNR), corrected for different statistics of the noise in the phantom compared with the background noise, was estimated for each image in the time sequence as

$$SNR = \frac{P_{center}}{\frac{1.53}{4} \sum_{i=1}^{4} SD_i} \qquad (12.1)$$

where P_{center} is the mean value of a 10×10 pixel area at the center of the image and SD_i is the mean standard deviation of the i-th of four 5×5 voxel ROIs areas at the image corners [13]. A total of 20 SNR estimates was computed for each sequence. As a second image quality index, the standard deviation of the image time course was estimated in each image sequence for a group of image volume elements, that is image voxel, in the central area of the phantom (15×15 voxel ROI). A total of 225 standard deviation estimates were computed for each image sequence. We choose the largest ROI size as possible, because to achieve a robust estimate relatively to the phantom size cross section at first and last slices. Image quality degradation with and without the glove was pinpointed by SNR values while time-domain standard deviations in the two conditions, were compared in order to determine if the device under investigation causes variances in the signal. As reference, two images (baseline I and II) were acquired in absence of the device in the scanner room. Then, an image sequence was acquired with the glove in the scanner room (80 cm far from the isocenter that is the bore center) in two conditions: while the glove was turned OFF, and while the glove was turned ON. Glove sensors were connected to the acquisition systems, located in the console room, by shielded cable passed through the waveguide of the scanner room Faraday shield. In order to test the differences between the SNR mean values, an unpaired two-sided t-test was chosen, given the small sample size. The significance level of the null hypothesis p, was set to 0.05, resulting in t critical values $t_c = \pm 2.024$. For the second quality index, a two sided z-test could be used given the large number or estimates. Setting the null hypothesis significance to 0.05, critical values for the statistic were $t_z = \pm 1.96$. When the results of the statistical test are outside the acceptance region, the null hypothesis of equality of means can be rejected, and it is thus possible to assert that the device caused significant artifacts in the images. The above statistical tests are valid when hypothesis of the Gaussian distribution of the SNR and SD estimates are satisfied. To verify the goodness of these hypotheses a Kolmogorov-Smirnov (KS) test (two sided KS, with a significance level equal to 0.05) was used. Moreover an F-test was performed to check the equality of variances in both cases.

12.3.2 fMRI Compatibility Test Results

KS test results confirmed the hypothesis that both SNR and time-domain standard deviation values are Gaussian-distributed for all samples and the F-test showed no significant differences between the variances in all examined cases. Finally, the compatibility test showed a mild effect of the system on image quality because changes in image quality are present as regards time-domain standard deviation in two slices, the first and the last slices, while others seem to be unaffected. No significant differences were found in SNR values across experimental conditions.

Effects of the Scanning System on the Sensing Glove

We also investigated the effect of the imaging system on our device, by acquiring the signal from the glove, located at the scanner bore entrance, both while the scanner was not operating and while the magnetic resonance system was scanning. Glove signals were acquired when an operator stretched the glove by holding the two edges of the elastic fabric during MR scanning: the contribution of the noise caused by scanning system was estimated using a power spectral density (PSD) analysis of the signal by means of a Welch-modified periodogram. From the analysis of the PSD of the acquired signal under different experimental conditions it is possible to assert that the scanner causes a relevant noise power on the acquired signal. However, the signal PSD is much larger than the noise for frequencies below 5 Hz, so a lowpass filtering operation (i.e. Butterworth order 12, cutoff frequency 5 Hz) can easily remove scanner noise.

fMRI Maps Reproducibility

Tests on volunteers performing a simple blocked design finger-tapping task were conducted. A comparison between results, obtained with and without the glove, was performed using a reproducibility measure [14], in order to test the effect of the wearable sensing glove on the fMRI maps. The reproducibility measure is defined as

$$R_{overlap}^{ij} = \frac{2V_{overlap}^{ij}}{V_i + V_j} \tag{12.2}$$

where V_i and V_j are the sizes of activated volumes in the ith and jth scan, respectively, while $V_{overlap}^{ij}$ is the size of the volume activated in both scans. The reproducibility measures estimated were above 63%. These values as compared to those found in literature [14, 15], indicate an acceptable similarity degree between the results obtained with and without wearing the sensing glove.

12.4 fMRI Study About Neural Correlates of Handshaking

This study is motivated by the need of improving our knowledge about interaction between human and social robots, that have been proposed for entertainment, therapy and as companions [16–18]. To realize an effective human-to-robot interaction, human-to-human interaction has to be characterized from the behavioral and cognitive point of view [19, 20]. In particular touch plays an important role in the communication between humans and robots [21], it improves both the perceived experience and the social presence in Virtual Reality scenarios [22, 23]. Research within HRI field is now pursuing an improvement in haptic devices both by developing new technological solutions and by studying the physiological basis of tactile perception. A relevant aspect is how social valence of tactile information is processed. The integration between vision and touch follows an information optimization rule, by weighting up the information coming from these modalities according to their reliability [24]. Moreover affective valence of visual information had been found to modulate tactile perception [25]. In this preliminary study we investigated the differences between human-to-human and human-to-robot interaction, by focusing on

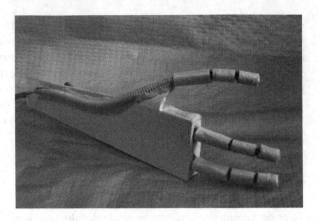

the relative importance of vision and touch in a handshaking task. This task is an important aspect of communication between two agents and has been studied from the behavioral point of view [26, 27] stressing the importance of context, gender and personality traits. The relevance of hand shape and softness on the quality of human-robot interaction [28]. Handshake has also been studied from the dynamic point of view, to this aim systems realizing human-to-robot handshaking have been developed [29–31].

12.4.1 Experimental Paradigm

Human and robotic, visual and haptic stimuli were realized and presented randomly to achieve congruent i.e. looking at a human face while shaking a human hand and incongruent i.e. looking at a robotic face while shaking a human hand tasks. Visual stimuli were static faces, projected on a back projection screen located at the foot of the scanner bed. Subjects could see the faces through the mirror located on the head coil, then they were asked to shake a human or robotic hand, while they were not aware about the hand identity. The robotic hand shown in Fig. 12.5 consisted in a plastic model of a hand with three fingers, comprising a thumb like finger to mimic human thumb opposition, and was actuated through plastic cables.

A full 2 × 2 fractional design was adopted (human face vs. robotic face x human hand vs. robotic hand). Subjects were presented with a human or robotic face, and after 1 s a beep cued them to shake the agent hand, presented in a standard position. Subjects were trained to minimize head movements, while actors were trained to minimize trial by trial variability in handshaking dynamics. A second beep, 3 s after the first one, was used to interrupt the tactile contact. The inter-stimulus interval was varied from 5 to 11 seconds and 18 handshaking trials were performed in each of the six runs. Three healthy, right handed, subjects (1 F, mean age 25 ± 2) were enrolled in the study. Functional images were acquired by using fMRI (GE-EPI sequence, FA = 90, TR = 2000 ms, TE = 40 ms, FOV = 24 cm, 24 slices, voxel size =

3.75 × 3.75 × 4 mm using a 1.5 T General Electric Scanner) while a 3D Spoiled Grass T1-weighted sequence was used to acquire anatomical information.

12.4.2 fMRI Data Analysis and Results

Data were first inspected for head movement effects and they were volume registered to a reference image. A 3D Gaussian kernel characterized by 8 mm Full Width Half Maximum parameter was used to further alleviate residual movement related artifacts, not compensated by volume registration step, and to reduce anatomical differences between subjects for successive group analysis. Functional and anatomical images, belonging to each subject, were mapped in the standard Talairach-Tournoux Atlas for group analysis. A multiple regression analysis at single subject level was used. Regressors timing was obtained from paradigm description and four regressors were created to describe expected BOLD changes elicited by handshake in the four conditions: two congruent conditions human-face-with-human-hand and robotic-face-with-robotic-hand, two incongruent conditions human-face-with-robotic-hand and robotic-face-with-human-hand. The relevance of haptic or visual stimuli was evaluated looking at the contrasts among congruent/incongruent conditions using a general linear test. A conjunction analysis was performed at group level to highlight the regions where contrasts were found significant ($p < 0.05$) across subjects.

AFNI was used to analyze data [32]. The results of a partial F-test about congruent human-to-human handshake, shown in Fig. 12.6, revealed a recruitment in all subjects of bilateral visual cortex, left precentral and postcentral gyri, left precuneus (Brodmann Area-BA-19), right superior temporal cortex right superior temporal cortex, bilateral middle temporal areas, left middle occipital gyrus, bilateral inferior frontal gyrus (BA44) and insula. In Fig. 12.7 results of the comparison between congruent human and congruent robot handshakes are shown. Congruent robot handshake recruited larger activated areas in right and left postcentral gyri, and right and left middle inferior and occipital gyri (BA19) with respect to congruent human handshake. No significant differences were highlighted between incongruent robot handshake (robotic face with human hand) and congruent human handshake (human face with human hand).

The contrasts between the handshaking with robotic and human hand, regardless face category, highlighted significant activations in bilateral middle occipital and left inferior occipital cortex. An open question is the role of the hand shape, because the robotic hand has a slimmer palm than the human hand. To verify this hypothesis the information acquired by the concurrent use of the sensing glove could be exploited as we will report in Sect. 12.4.3.

12.4.3 Glove/fMRI Data Integration

To verify the benefits of task execution monitoring using the sensing glove, we asked to our volunteers to wear the sensing glove while they were executing the previously

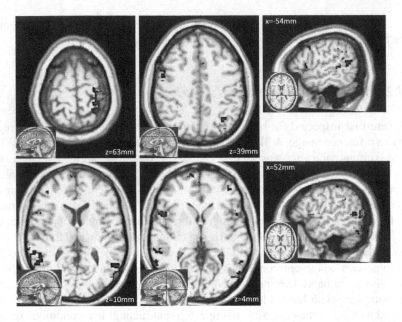

Fig. 12.6 Conjunction analysis results at group level pertaining congruent human handshake. Areas that were found activated in all subjects, are shown in Talairach-Tournoux Atlas. Image orientation is RAI. Reproduced with permission from [9] © 2010 IEEE

described handshaking fMRI task. Results obtained using the actual task timings as extracted from data glove, were compared with those using only paradigm information, as described in Sect. 12.4.2. While the difference between expected and actual movement onset times on average is only 46 ms, a standard deviation of 304 ms was observed. We have to point out that a negative difference is due to the movement onset anticipating the starting movement cue, expected 1 s after visual stimulus presentation. Moreover a different shake duration was observed between human and robot handshake (2.2 ± 0.49 s and 3.2 ± 0.27 s respectively). The two different General Linear Models, obtained with expected and actual timings, were compared looking at the coefficient of multiple determination R^2. This coefficient ranges from 0 to 1 and it is referring to the proportion of data variance with respect to the baseline model, i.e. the model excluding the regressors of interests as those task related, explained by the full model. In Fig. 12.8 the histogram of the distribution of R^2 statistic differences as obtained from the two models, using and not using the glove, is shown. A slight increase using the data glove can be observed. Moreover a 5% increase in the number of significant ($p < 10^{-5}$) voxels was observed.

To verify the hypothesis that differences between brain activity while handshaking a human hand and while handshaking a robot hand were not related to hand shape, trial by trial changes in fingers joint angles were estimated. In particular we extracted from the data glove, the maximum finger joint angles during the closing phase. We observed significant smaller angles for thumb and index/middle fingers and larger ring/little fingers angles while handshaking a robot hand than during hu-

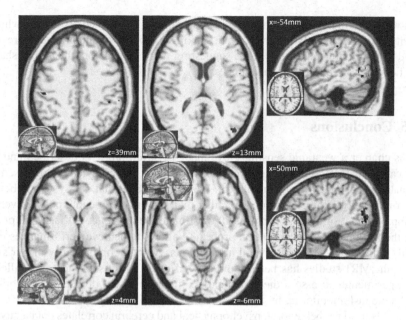

Fig. 12.7 Conjunction analysis results at group level, pertaining contrast between congruent-robot and congruent-human handshake tasks. Positive contrast results were observed, resulting in larger activations for congruent-robot with respect to congruent-human handshake. Images are in the Talairach-Tournoux Atlas and orientation is RAI. Reproduced with permission from [9] © 2010 IEEE

Fig. 12.8 The distribution of R^2 statistic difference between models obtained using sensing glove and paradigm information

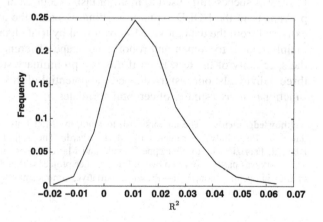

man handshaking. This information was used as auxiliary behavioral information (ABI) in a regression model to look at brain areas whose activities are modulated by joint angle changes. Significant correlations between brain activity changes and ABI were observed in bilateral SMA and ipsilateral primary motor area (BA6), during the congruent robotic condition—when subjects where looking at a robot face and handshaking a robot hand. A negative correlation with ABI was found in bilateral posterior parietal cortex (BA7), related to visuo-motor coordination, while volunteers were looking at a human face and handshaking a robot hand. The areas whose activity was significantly different when shaking a robot or a human hand were not found to be modulated by this ABI, suggesting that other factors than hand shape may be responsible for these differences.

12.5 Conclusions

In this chapter a wearable sensing glove made of strain fabric is described and its application in fMRI studies has been discussed. The glove device can be used to estimate in real time finger joint angles. We demonstrated that our system is better performing during the closing phase, rather than the fingers opening phase. Even if an improvement of the algorithm used to estimate the angles could face this problem, the fast and simple calibration procedure allows to achieve performances that could fulfill the specifics of several applications. A good compatibility of our system with fMRI studies has been demonstrated using a statistical test, that allows the experimenter to assess the statistical significance of the device compatibility. The haptic task monitoring by the sensory glove could aid in improving the statistical models used for behavioral, psychophysical and cerebral correlates data analysis and in verifying task execution. The possibility of estimating other measures of hand dynamics could then be exploited to increase the specificity of the hypothesis being tested. In direction, in this chapter a preliminary study about differences of neural correlates of human-to-human and human-to-robot handshaking is proposed. The glove was successfully used to monitor task execution and we demonstrated an improvement in the BOLD signal modelling by using the actual timings information extracted from the data glove. Moreover trial by trial changes in joint finger angles, mainly related to human and robot hand shape differences, were used to improve the specificity of the test. Even if this is a preliminary study and we enrolled only three individuals, our results showed the potential of this system as integrated with functional in vivo studies of cerebral correlates.

Acknowledgements The authors wish to thank the MRI Laboratory at G. Monasterio Foundation in Pisa, Italy, coordinated by M. Lombardi. The authors are grateful to Eng. Paolo Bianchi, Fabrizio Cutolo, Giuseppe Zupone and Mario Tesconi for contributions to this work. This work was partly supported by the ImmerSence project within the 6th Framework Programme of the European Union, FET—Presence Initiative, contract number IST-2006-027141, see also www.immersence.info.

References

1. Lorussi, F., Scilingo, E.P., Tesconi, M., Tognetti, A., De Rossi, D.: Strain sensing fabric for hand posture and gesture monitoring. IEEE Trans. Inf. Technol. Biomed. **9**(3), 372–381 (2005)
2. Hidler, J., Hodics, T., Xu, B., Dobkin, B., Cohen, L.G.: MR compatible force sensing system for real-time monitoring of wrist moments during fMRI testing. J. Neurosci. Methods **155**(2), 300–307 (2006)

3. Liu, J., Dai, T., Elster, T., Sahgal, V., Brown, R., Yue, G.: Simultaneous measurement of human joint force, surface electromyograms, and functional MRI-measured brain activation. J. Neurosci. Methods **101**(1), 49–57 (2000)
4. Reithler, J., Reithler, H., van den Boogert, E., Goebel, R., van Mier, H.: Resistance-based high resolution recording of predefined 2-dimensional pen trajectories in an fMRI setting. J. Neurosci. Methods **152**(1–2), 10–17 (2006)
5. Tada, M., Kanade, T.: An MR-compatible optical force sensor for human function modeling. In: Medical Image Computing and Computer-Assisted Intervention—MICCAI 2004, pp. 129–136 (2004)
6. James, G.A., He, G., Liu, Y.: A full-size MRI-compatible keyboard response system. NeuroImage **25**(1), 328–331 (2005)
7. Gassert, R., Moser, R., Burdet, E., Bleuler, H.: MRI/fMRI-compatible robotic system with force feedback for interaction with human motion. IEEE/ASME Trans. Mechatron. **11**(2), 216–224 (2006)
8. Vanello, N., Hartwig, V., Tesconi, M., Ricciardi, E., Tognetti, A., Zupone, G., Gassert, R., Chapuis, D., Sgambelluri, N., Scilingo, E.P., et al.: Sensing glove for brain studies: Design and assessment of its compatibility for fMRI with a robust test. IEEE/ASME Trans. Mechatron. **13**(3), 345–354 (2008)
9. Vanello, N., Bonino, D., Ricciardi, E., Tesconi, M., Scilingo, E., Hartwig, V., Tognetti, A., Zupone, G., Cutolo, F., Giovannetti, G., et al.: Neural correlates of human-robot handshaking. In: RO-MAN, pp. 555–561. IEEE, New York (2010)
10. Skopec, M.: A primer on medical device interactions with magnetic resonance imaging systems, Feb. 4, 1997, CDRH magnetic resonance working group. US Department of Health and Human Services, Food and Drug Administration. Center for Devices and Radiological Health, Updated May 23 17 (1997)
11. Schenck, J.F.: Safety of strong, static magnetic fields. J. Magn. Reson. Imaging **12**(1), 2–19 (2000)
12. Lüdeke, K., Röschmann, P., Tischler, R.: Susceptibility artefacts in NMR imaging. J. Magn. Reson. Imaging **3**(4), 329 (1985)
13. Sijbers, J., Den Dekker, A., Van Audekerke, J., Verhoye, M., Van Dyck, D.: Estimation of the noise in magnitude MR images. J. Magn. Reson. Imaging **16**(1), 87–90 (1998)
14. Rombouts, S., Barkhof, F., Hoogenraad, F., Sprenger, M., Valk, J., Scheltens, P.: Test-retest analysis with functional MR of the activated area in the human visual cortex. Am. J. Neuroradiol. **18**(7), 1317 (1997)
15. Machielsen, W.C.M., Rombouts, S.A.R.B., Barkhof, F., Scheltens, P., Witter, M.P.: FMRI of visual encoding: reproducibility of activation. Hum. Brain Mapp. **9**(3), 156–164 (2000)
16. Carelli, L., Gaggioli, A., Pioggia, G., De Rossi, F., Riva, G.: Affective robot for elderly assistance. Stud. Health Technol. Inform. **144**, 44 (2009)
17. Pioggia, G., Igliozzi, R., Ferro, M., Ahluwalia, A., Muratori, F., De Rossi, D.: An android for enhancing social skills and emotion recognition in people with autism. IEEE Trans. Neural Syst. Rehabil. Eng. **13**(4), 507–515 (2005)
18. Bird, G., Leighton, J., Press, C., Heyes, C.: Intact automatic imitation of human and robot actions in autism spectrum disorders. Proc. R. Soc. Lond. B, Biol. Sci. **274**(1628), 3027 (2007)
19. Sidner, C.L., Lee, C., Kidd, C.D., Lesh, N., Rich, C.: Explorations in engagement for humans and robots. Artif. Intell. **166**(1–2), 140–164 (2005)
20. Nehaniv, C.L., Dautenhahn, K., Kubacki, J., Haegele, M., Parlitz, C., Alami, R.: A methodological approach relating the classification of gesture to identification of human intent in the context of human-robot interaction. In: IEEE International Workshop on Robot and Human Interactive Communication. ROMAN 2005, pp. 371–377. IEEE, New York (2005)
21. Lee, K.M., Jung, Y., Kim, J., Kim, S.R.: Are physically embodied social agents better than disembodied social agents?: The effects of physical embodiment, tactile interaction, and people's loneliness in human-robot interaction. Int. J. Hum.-Comput. Stud. **64**(10), 962–973 (2006)
22. Basdogan, C., Ho, C.H., Srinivasan, M.A., Slater, M.: An experimental study on the role of touch in shared virtual environments. ACM Trans. Comput.-Hum. Interact. **7**(4), 443–460 (2000)

23. Giannopoulos, E., Eslava, V., Oyarzabal, M., Hierro, T., González, L., Ferre, M., Slater, M.: The effect of haptic feedback on basic social interaction within shared virtual environments. In: Haptics: Perception, Devices and Scenarios, pp. 301–307 (2008)
24. Ernst, M.O., Banks, M.S.: Humans integrate visual and haptic information in a statistically optimal fashion. Nature 415(6870), 429–433 (2002)
25. Montoya, P., Sitges, C.: Affective modulation of somatosensory-evoked potentials elicited by tactile stimulation. Brain Res. 1068(1), 205–212 (2006)
26. Chaplin, W.F., Phillips, J.B., Brown, J.D., Clanton, N.R., Stein, J.L.: Handshaking, gender, personality, and first impressions. J. Pers. Soc. Psychol. 79(1), 110 (2000)
27. Åtröm, J., Thorell, L.H., Holmlund, U., d'Elia, G.: Handshaking, personality, and psychopathology in psychiatric patients: A reliability and correlational study. Perceptual and motor skills (1993)
28. Zecca, M., Endo, N., Itoh, K., Imanishi, K., Saito, M., Nanba, N., Takanobu, H., Takanishi, A.: On the development of the bioinstrumentation system WB-1R for the evaluation of human-robot interaction-head and hands motion capture systems. In: 2007 IEEE/ASME International Conference on Advanced Intelligent Mechatronics, pp. 1–6. IEEE, New York (2007)
29. Wang, Z., Peer, A., Buss, M.: An HMM approach to realistic haptic human-robot interaction. In: Third Joint EuroHaptics Conference, 2009 and Symposium on Haptic Interfaces for Virtual Environment and Teleoperator Systems. World Haptics 2009, pp. 374–379. IEEE, New York (2009)
30. Sato, T., Hashimoto, M., Tsukahara, M.: Synchronization based control using online design of dynamics and its application to human-robot interaction. In: IEEE International Conference on Robotics and Biomimetics. ROBIO 2007, pp. 652–657. IEEE, New York (2007)
31. Yamato, Y., Jindai, M., Watanabe, T.: Development of a shake-motion leading model for human-robot handshaking. In: SICE Annual Conference, pp. 502–507. IEEE, New York (2008)
32. Cox, R.W., et al.: AFNI: software for analysis and visualization of functional magnetic resonance neuroimages. Comput. Biomed. Res. 29(3), 162–173 (1996)

Chapter 13
Improving Human-Computer Cooperation Through Haptic Role Exchange and Negotiation

Ayse Kucukyilmaz, Salih Ozgur Oguz, Tevfik Metin Sezgin,
and Cagatay Basdogan

Abstract Even though in many systems, computers have been programmed to share control with human operators in order to increase task performance, the interaction in such systems is still artificial when compared to natural human-human cooperation. In complex tasks, cooperating human partners may have their own agendas and take initiatives during the task. Such initiatives contribute to a richer interaction between cooperating parties, yet little research exists on how this can be established between a human and a computer. In a cooperation involving haptics, the coupling between the human and the computer should be defined such that the computer can understand the intentions of the human operator and respond accordingly. We believe that this will make the haptic interactions between the human and the computer more natural and human-like. In this regard, we suggest (1) a role exchange mechanism that is activated based on the magnitude of the force applied by the cooperating parties and (2) a negotiation model that enables more human-like coupling between the cooperating parties. We argue that when presented through the haptic channel, the proposed role exchange mechanism and the negotiation model serve to communicate the cooperating parties dynamically, naturally, and seamlessly, in addition to improving the task efficiency of the user. In this chapter, we explore how human-computer cooperation can be improved using a role-exchange mechanism and a haptic negotiation framework. We also discuss the use of haptic negotiation in assigning different behaviors to the computer; and the effectiveness of visual and haptic cues in conveying negotiation-related complex affective states. Throughout this chapter, we will adopt a broad terminology and speak of cooperative systems, in which both parties take some part in control, as shared control

A. Kucukyilmaz (✉) · S.O. Oguz · T.M. Sezgin · C. Basdogan
Koc University, Istanbul 34450, Turkey
e-mail: akucukyilmaz@ku.edu.tr

S.O. Oguz
e-mail: ooguz@cs.ubc.ca

T.M. Sezgin
e-mail: mtsezgin@ku.edu.tr

C. Basdogan
e-mail: cbasdogan@ku.edu.tr

A. Peer, C.D. Giachritsis (eds.), *Immersive Multimodal Interactive Presence*,
Springer Series on Touch and Haptic Systems,
DOI 10.1007/978-1-4471-2754-3_13, © Springer-Verlag London Limited 2012

schemes, but the term "control" is merely used to address the partners' manipulation capacities on the task.

13.1 Introduction

Haptic cooperation involves the interaction of two parties through the sense of touch. Such interaction is often implemented between humans and between humans and machines in order to create a better sense of immersion. Shared control between a human and a computer evolved from the idea of supervisory control since its emergence in early 1960s [1]. Supervisory control has been typically used in teleoperation tasks which are difficult to automate. It allows the human operator (the master) to assume the role of the supervisor or the decision maker, while the robot operated by the computer (the slave) executes the task. However, human-human cooperation is far richer and more complex than this scheme, since the exchange of the roles and the control of the parties on the task is dynamic and the intentions are conveyed through different sensory modalities during the execution of the task. Moreover, negotiation is a significant component of interaction in human-human cooperation.

The shared control systems available today for human-computer cooperation possess only a subset of these features (see the review in [2]). Hence, as cooperative tasks get more complex, such schemes fall short in providing a natural interaction that resembles human-human communication. In order to alleviate this deficiency, a mechanism, where both parties can be employed with different levels of control during the task, is needed. In the last decade, interactive man-machine systems with adjustable autonomy have been developed. Adjustable autonomy is implemented to make teamwork more effective in interacting with remote robots by interfacing the user with a robot at variable autonomy levels [3, 4]. These autonomy levels imply different role definitions for human and computer partners.

Lately, the notion of exchanging roles also emerged in the context of haptic collaboration. Several groups examined role exchange in human-human collaboration. Nudehi et al. [5] developed a haptic interface for training in minimally invasive surgery. The interface allowed to shift the "control authority" shared between two collaborating human operators, based on the difference of their actions. Reed and Peshkin [6] examined dyadic interaction of two human operators in a 1 DOF target acquisition task and observed different specialization behaviors of partners such as accelerators and decelerators. However, they did not comment on the possible reasons or the scheduling of this specialization. Stefanov et al. [7] proposed executor and conductor roles for human-human haptic interaction. In their framework, the conductor assumed the role of deciding on the system's immediate actions and expressing his/her intentions via haptic signals so that the executor can perform these actions. They proposed a model for role exchange using the velocity and the interaction force. This system is especially interesting in a sense that the parties are required to communicate only through the haptic channel, i.e. the conductor is assumed to express his/her intention by applying larger forces. Also in this work, they examined the phases of interaction that lead to different role distributions. In a recent paper, Groten et al. [8] investigated the effect of haptic interaction in different shared decision situations in human-human cooperation, where an operator can choose to agree/disagree with the intention of his/her partner or to remain passive and obey his/her partner in a path following task. They observed that when operators have disagreement in their actions, the amount of physical effort is increased (interpreted as additional negotiation effort) and performance is decreased. They also found that the existence of haptic feedback further increases the physical effort but improves the performance. The findings of this study is in conflict with their previous work in [9], where haptics increased the effort but provided no performance gains. Even though they conclude that this result might stem from the fact that the former task included no negotiation, their findings are conclusive.

Even though the studies mentioned above presented very important observations regarding human-human interaction, only few groups focused on role definitions and exchange in human-computer interaction involving haptics. Evrard et al. [10] studied role exchange in a symmetric dyadic task where a human interacts with a computer partner through an object. They allowed the operators to switch between leader and follower roles during the task. In order to describe role exchange for collaborative interaction, they used two functions to model different interaction behaviors. However, they failed to implement a user-centric and dynamic negotiation mechanism to handle the interaction between a human and a computer. Oguz et al. [11] came up with a haptic negotiation framework to be used in dynamic tasks where a human negotiates with a computer. This framework defined implicit roles in terms of the parties' control levels, and dynamically realized role exchanges between the cooperating parties. They defined two extremes for identifying different roles for sharing control: user dominant and computer dominant control levels. Their haptic negotiation model allowed dynamic and user specific communication with the computer. The role exchanges were performed regarding the magnitude of the force applied by the cooperating parties. The results of this study indicated that the suggested negotiation model introduced a personal and subjectively pleasing interaction model and offered a tradeoff between task accuracy and effort. However, even though they claimed that the framework was built to enable negotiation, their task was strictly collaborative, where the communication addressed a decision making process between the parties rather than negotiation. Additionally, since the users were not informed on the nature of the task that they were performing, the evaluation of the utility of the role exchange mechanism was not feasible.

Table 13.1 presents a comparison of selected shared control studies that implemented roles as explained within this section. The features of the haptic board game and the haptic negotiation game (explained respectively in Sects. 13.3.1 and 13.3.2) developed by us are included in the last two rows of the table. Both games use the negotiation framework suggested in [11] and mainly investigates the effectiveness of haptic cues on communication in collaborative or conflicting situations.

Upon close inspection of the table, we observe that half of the tasks in question focus only on human-human interaction, while the remaining realize human-computer interaction. An immediate examination shows that in all tasks, the parties have a common goal they want to optimize, but only in two (Groten et al. [8] and

Table 13.1 A comparison of several shared control tasks implemented in the recent literature

	Human-Computer Interaction	Simultaneous Shared Control	Sequential Shared Control	Common Goal	Separate Agendas	Separate Roles	Dynamic Role Exchange
Reed & Peshkin [6]	×	✓	×	✓	×	✓	×
Stefanov et al. [7]	×	✓	×	✓	×	✓	✓
Groten et al. [8]	×	✓	×	✓	✓	✓	✓
Evrard et al. [10]	✓	✓	✓	✓	×	✓	×
Kucukyilmaz et al. [12]	✓	×	✓	✓	×	✓	✓
Oguz et al. [13]	✓	✓	×	✓	✓	✓	×

Oguz et al. [13]), they have separate agendas. Also it can be seen in columns 2 and 3 that only Evrard et al. [10] implemented a shared control scheme that is both simultaneous and sequential,[1] however a dynamic role exchange between the cooperating parties has not been considered as it is done in Kucukyilmaz et al. [12].

Specifically, this chapter aims to present a perspective to build more natural shared control systems for human-computer cooperation involving haptics. We suggest that a cooperative haptic system will be more effective if the human's and the computer's levels of control are dynamically updated during the execution of the task. These control levels define states for the system, in which the computer's control leads or follows the human's actions. In such a system, a state transition can occur at certain times if we determine the user's intention for gaining/relinquishing control. Specifically, with these state transitions we assign certain roles to the human and the computer. Also, we believe that only by letting the computer have a different agenda than that of the user's, we can make emotional characteristics of touch transferred to the human participant.

In this sense, this chapter gives details on the proposed haptic negotiation model and presents two applications we have developed earlier to investigate how this model can be used to realize natural collaboration and adapted to systems where the computer is programmed to display different behaviors when interacting with the user. In the next section, we present the haptic negotiation model we introduced in [11] in its general form. In Sects. 13.3.1 and 13.3.2, we present two games; one for demonstrating the effectiveness of the proposed haptic role exchange mechanism [12], and the other for demonstrating the use of the proposed haptic negoti-

---

[1] A general categorization of shared haptic interaction, which is similar to Sheridan's classification, talks about "*simultaneous*" versus "*sequential*" haptic manipulation classes [14]. In simultaneous haptic manipulation, both parties can actively control the task concurrently, whereas in sequential manipulation, they take turns in control.

Fig. 13.1 The general
negotiation model

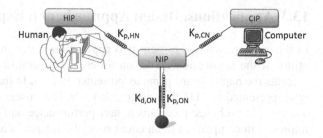

ation model to assign different behaviors to the computer [13]. Finally Sect. 13.4
summarizes our results and Sect. 13.5 presents conclusions.

13.2 Haptic Negotiation Model

The proposed negotiation model in its general form is sketched in Fig. 13.1. This
model is developed to let the human and the computer participants interact to move
a virtual object via a spring-damper system governed by three interface points that
act like massless particles. The human and the computer basically interface with the
system through individual interface points, which are labeled as HIP (user's Haptic
Interface Point) and CIP (Computer's Interface Point) in the figure. These two points
are interconnected at a negotiated interface point (NIP), which directly pulls the object towards itself, so that the control is shared between the parties. This system
facilitates the operation of assigning different roles (i.e. control levels) to the parties by changing the stiffness and damping coefficients of the system (represented
respectively by K_p and K_d values in the figure). $K_{p,ON}$ affects how much the manipulated object is diverted from the negotiated path. The human and the computer
are granted different levels of control on the game by changing the stiffness coefficients between CIP and NIP ($K_{p,CN}$) and between HIP and NIP ($K_{p,HN}$), as
depicted on the figure. If $K_{p,CN}$ and $K_{p,HN}$ have equal value, the computer and
the user will have equal control on the game. The computer will be the dominant
actor within the game if $K_{p,CN}$ has a larger value, and vice versa, the user will be
dominant if $K_{p,HN}$ is larger.

The negotiation model can be used to realize decision making and negotiation.
By dynamically shifting the stiffness of the system in favor of the user or the computer, we can also realize a smooth and natural transition between different roles
for the parties. Also programming the computer to execute different behaviors is
straightforward because of the disjoint nature of the model and the operations of the
parties. This way, a richer interaction can be achieved between the human and the
computer. Since a human being will possess a variety of different behaviors and realize many negotiation processes in a dynamic task, using such a model is not only
beneficial, but also necessary to come up with a natural human-like communication.

13.3 Applications, Design Approach, and Experiments

In this section, we will present two applications through which we displayed the utility of the negotiation model on different cooperative schemes. Section 13.3.1 presents the haptic board game as presented in [12]. In that study, we extended the work presented in [11] to illustrate that when the users are instructed on how to use the role exchange mechanism, task performance and efficiency of the user are improved in comparison to an equal control guidance scheme. We also augmented the system with informative visual and vibrotactile cues to display the interaction state to the user. We showed that such cues improve the user's sense of collaboration with the computer, and increase the interaction level of the task.

Section 13.3.2 introduces the haptic negotiation game as presented in [13]. In the haptic negotiation game, the computer is programmed to display different behaviors, competitive, concessive, and negotiative; and the extent to which people attributed those behaviors to the computer is investigated. We observed that the users can more successfully identify these behaviors with the help of haptic feedback. Moreover, the negotiative behavior of the computer generates higher combined utility for the cooperating parties than that of the concessive and competitive behaviors. However, there is a trade-off between the effort made by the user and the achieved utility in negotiation.

13.3.1 Haptic Board Game

In order to create a collaborative virtual environment, we implemented the negotiation model and the role exchange mechanism on top of a haptic guidance system in a complex and highly dynamic interactive task [12]. Our task is a simple board game in a virtual environment, in which the user controls the position of a ball with the help of a SensAble Technologies PHANTOM® Premium haptic device to hit targets on a flat board. The board is tilted about x and z axes as a result of the movement of the ball, and the user is given haptic feedback due to the dynamics of the game.

Rules and Design Considerations

Figure 13.2 presents a screenshot of the board game. The board employs a ball and four cylinders. Each cylinder is imprisoned in between two pits which diagonally extend towards the center of the board. The aim of the game is to move the ball to hit the target cylinder and wait on it to the count of 10. The users are asked to perform the task in minimum time and to avoid falling into the pits lying on the left and right side of each cylinder. The pits serve as penalty regions within the game. If the ball falls in a pit, a fault occurs, all acquired targets are canceled, and the timer runs faster as long as the ball is in the pit. Hence, the faults impair the final score of the users. To leave the pit, the user should move the ball towards the entrance of the pit.

Fig. 13.2 A screenshot of the Haptic Board Game

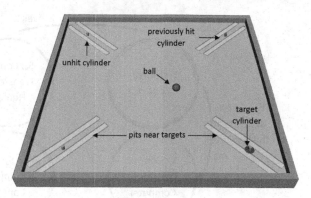

Fig. 13.3 The physics-based model for the haptic board game

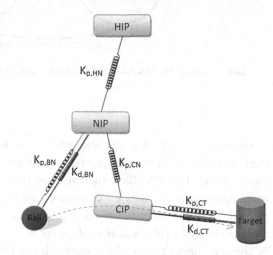

Since precision is required to avoid pits, we assumed that the users would benefit from computer guidance especially when they are within the pits. A preliminary investigation of the users' role exchange patterns confirmed this assumption, such that the users indeed requested computer guidance in these regions.

At the start of each game, all but the first target cylinder are colored in grey. The target cylinder is highlighted in green color. When the user hits the target, it turns red and a new target is automatically selected and colored accordingly. In Fig. 13.2, the cylinder in the upper right corner of the board is previously hit, whereas the target cylinder lies at the lower right corner. The remaining two cylinders are unhit.

The negotiation model used in the haptic board game is sketched in Fig. 13.3. A spring and a damper is added to the system to implement a PD (Proportional-Derivative) control algorithm [15] that lets the computer move the ball towards the targets when it is granted control. The user is presented with force feedback due to the tilt of the board and the guidance provided by the computer if available.

Since the underlying system is highly dynamic, a sophisticated mechanism is required to realize the role exchanges. Hence, we implemented a dynamic role exchange mechanism on top of the board game to allow computer guidance. This

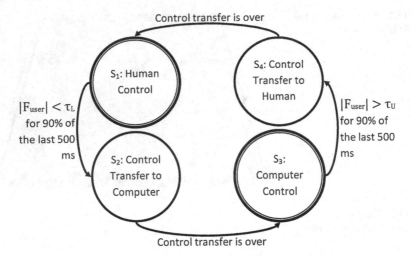

Fig. 13.4 The state diagram defining the role exchange policy

mechanism allows the user to communicate with the computer dynamically through the haptic channel to realize role exchange using a four-state finite state machine as shown in Fig. 13.4. The finite state machine realizes a smooth transition during the exchange of roles. Initially the computer determines thresholds based on the user's average force profile to assign roles to each entity. It is assumed that the user is trying to initiate a role exchange whenever the magnitude of the force (s)he applies is above an upper threshold or below a lower threshold for over a predetermined amount of time. These threshold values are calculated using the average force profile of the user, i.e. the user's average forces and the standard deviation of these forces, and are adaptively updated during the course of the game. In Fig. 13.4, F_{user} denotes the instantaneous force applied by the user. $\tau_{Th,L}$ and $\tau_{Th,U}$ refer to the personalized lower and upper threshold values for initiating state transitions.

The states of the finite state machine also define the interaction states within the system. Initially the system is in the state S1 (*human control*), in which the user controls the ball on his/her own. When the user tries to get assistance, we assume that (s)he will stay below the calculated lower threshold value for more than 90% of the last 500 milliseconds. In this case, the system enters the state S2 (*control transfer to computer*), in which the computer gradually takes control until it gains full control of the ball. The state will automatically change from the state S2 to S3 (*computer control*) after 1000 milliseconds. A similar scenario is also valid when the user decides to take over control, so that a series of state transitions will occur from the state S3 to the state S4 (*control transfer to human*) and the state S1 in succession.

Fig. 13.5 Two different configurations for the role indication icons

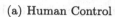

            (a) Human Control   (b) Computer Control

Augmenting Sensory Elements

We also augmented our system with additional visual and vibrotactile cues to display the control state to the users. These cues are meant to make the role exchange process more visible, hence we chose to present all cues simultaneously, so that by acquiring the same information through multiple resources, the information processing is facilitated more effectively for the users.

Visual Cues   Two icons are displayed above the board to represent the control levels possessed by the human and the computer as exemplified in Fig. 13.5. The icons grow and shrink gradually to simulate the state transitions that take place when parties exchange roles. The larger icon demonstrates the party that holds a greater level of control on the movement of the ball. For instance, in Fig. 13.5(a), the system is in human control state since the icon for the human (on the right) has maximum size. Similarly, in Fig. 13.5(b), the computer holds control because its icon (on the left) is larger.

Vibrotactile Cues   Two different vibrotactile cues are used to signal the state transitions within the system:

Buzzing is presented to the user to signal the initiation of a role exchange. We implemented buzzing as a high frequency vibration (100 Hz) presented through the haptic device.

Tremor is displayed to the user to imitate the hand tremor of the collaborating entity that the user is working with. The tremor effect is generated by varying the frequency (8–12 Hz) and amplitude of the vibrotactile signal when the computer has full control.

Experimental Design

In order to investigate the effect of role exchange on collaboration, we tested four conditions with the haptic board game:

No Guidance (NG): The user plays the game without any computer assistance to control the ball position on the board.

Equal Control (EC): The user and the computer share control equally at all times to move the ball.

Role Exchange (RE): At any point during the game, the user can hand/take over the control of the ball to/from the computer, by altering the forces (s)he applies through the haptic device.

VisuoHaptic Cues (VHC): As in RE condition, the user decides to hand/take over the control of the ball to/from the computer. Also, role indication icons, buzzing, and tremor will be displayed to the user to inform her/him about the state of control during the game.

30 subjects (9 female and 21 male), aged between 21–28, participated in our study. A within subjects experiment, in which each subject experimented all four conditions in a single day, is conducted. An experiment consisted of an evaluation and a post-evaluation session. In the evaluation session, the subjects played the haptic board game 5 times successively in each condition. We presented NG condition at the beginning of each experiment as a baseline condition. Afterwards, we presented the guidance conditions EC, RE, and VHC. However, in order to avoid learning effects, the conditions were permuted and 5 subjects were tested in each of the six permutations of three guidance conditions. The subjects were given detailed information about the conditions and instructed on how to use the underlying role exchange mechanism when applicable. In order to avoid any perceptual biases, instead of naming the conditions, the subjects were shown aliases such as Game A, B, and C. After the evaluation session, the post-evaluation session was started, in which the subjects played under each condition once (i.e. one trial only).

Measures

Quantitative Measures In each trial, we collected the task completion time and the number of faults the user makes, as measures of task performance. Task completion time is recorded in milliseconds. The number of faults is the count of times that the user falls into a pit during a trial. A penalized completion time is computed such that in case of faults, i.e. when the user falls into a pit, the timer is iterated 10 times faster. Hence, the penalized completion time is greater than completion time unless the user completes the game without any faults.

Additionally, we calculated the energy spent by the user during the trial to represent the physical effort (s)he spends in competing the task. This energy is calculated by the dot product of the displacement of HIP and the force exerted by the spring located between NIP and HIP as:

$$energy = \int_{P_H} |F_{HIP} \cdot dx_{HIP}|,$$

where P_H is the path traversed by HIP during the trial, F_{HIP} is the force exerted by the spring between HIP and NIP, and x_{HIP} is HIP's position.

Along with the user's energy requirement, we calculated the work done by him/her on the ball. The work done on the ball is computed regarding the displacement of the ball and the force acting on the ball due to the user's actions:

$$work = \int_{P_B} |F_{HIP} \cdot dx_{ball}|,$$

where P_B is the path traversed by the ball during the trial and x_{ball} equals to the ball's position. We presume that the work done on the ball is a measure of the user's contribution to the task. We suggest that, in an efficient collaborative system, the user should work in harmony with the computer. This way, the user's effort will be spent for completing the task, not in resisting the operation of the computer. Motivated with this conception, we define an efficiency measure as the ratio of the work done by the user on the ball to the energy she/he spends. Our efficiency measure is calculated as:

$$efficiency = \frac{work}{energy},$$

Upon closer inspection, we can see that, as expected, the efficiency is maximized when the user does a large amount of work on the ball with a small effort.

Subjective Measures   At the end of each experiment, we asked the subjects to fill a questionnaire to comment on their experiences in the guidance conditions (EC, RE, VHC). For the questionnaire design, we used the technique that Basdogan et al. used previously for investigating haptic collaboration in shared virtual environments [14]. A 7-point Likert scale was used for the answers; and questions were rephrased and scattered randomly within the questionnaire. For evaluation, the averages of the questions, which are rephrased, are used. Some of the variables investigated in the questionnaire are:

- *Collaboration:* 2 questions investigated whether the subjects had a sense of collaborating with the computer or not.
- *Interaction:* 5 questions explored the level of interaction that the subjects experienced during the task.
- *Comfort and pleasure:* 4 questions investigated how comfortable and pleasurable the task was.
- *Trust:* 2 questions investigated whether the users trusted their computer partner on controlling the ball or not.
- *Role exchange visibility:* A single question explored whether or not the users observed the functionality of the role exchange processes during the task.

13.3.2 Haptic Negotiation Game

Our second task requires a human to play negotiation game with the computer. In this game, the computer is equipped with the ability to display different negotiation behaviors to cooperate with the user.

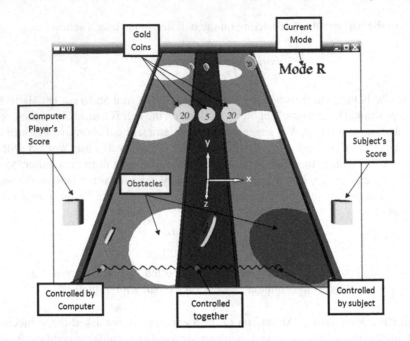

Fig. 13.6 A screenshot of the Haptic Negotiation Game

Rules and Design Considerations

On the screen, the user sees a road divided into 3 lanes (Fig. 13.6). On the left-hand side, the computer player controls CIP (the green ball) to avoid obstacles and collects coins to increase its score. Likewise, on the right-hand side, the user controls HIP (the blue ball) to avoid obstacles and collect coins to increase her own score. The middle lane also has a coin which can be collected by the red ball—referred to as "the Ball" in the rest of the text. The position of the Ball is controlled by the subject and the computer together. As the players control their respecting interface points (IPs) in the horizontal plane, the obstacles scroll down the screen with a constant velocity.

Separate scores are calculated for the user and the computer. The user's score is calculated by summing up the values of coins that (s)he collects from the middle lane and from his/her own lane. The computer's score is calculated by summing up values of coins that it collects from the middle lane and from its own lane. The scores for each player can be seen on the left and right margins of the screen represented as bars that filled with coins collected by the users. When a coin is collected by the Ball, it gets awarded to both players and its value is added to the scores of both dyads. Since the Ball is controlled together by both players, they need to collaborate in order to ensure that the Ball collects the coins in the middle lane. However, certain layouts of the obstacles in the computer and human players' lanes may cause conflicting situations where collecting the coin in the middle lane necessarily requires one of the players to hit an obstacle, hence miss the coin in his/her/its own

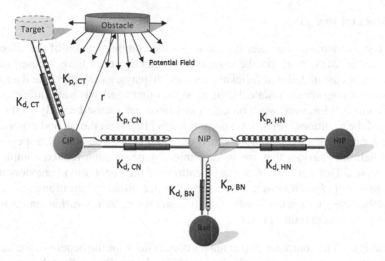

Fig. 13.7 The physics-based model for the haptic negotiation game. All interface points and the ball move on the same horizontal line, but for clarity, the ball is illustrated as if it falls behind the negotiated interface point

lane.[2] By design, players can collect coins in their lanes, but they need to cooperate in order to have the coin in the middle be collected by the Ball if they want to obtain the score of the middle coin. Otherwise, their movements conflict with each other, and the Ball fails to collect the middle coin. In other words, this conflicting situation requires one of the players to concede and thus help the other to acquire his/her/its own coin.

The general negotiation model is modified as in Fig. 13.7. Apart from constraining the movements of HIP and CIP to the horizontal line, CIP is made to be affected also by external forces due to the potential field around the obstacles in addition to the attractive forces moving CIP to the predefined target points along the path. The potential field exerts a force inversely proportional to r^2, where r is the distance between the CIP and the obstacle, hence is used to assist the computer in avoiding the obstacles and reaching the predefined target points. It can be turned on and off according to the behavior assigned to the computer player.

In order to better convey conflicting and cooperative behavior, we apply forces to the haptic device held by the user. We use the haptic channel for conveying the negotiation dynamics. Hence, the users feel the forces due to the Ball's deviation from the initial point, which is the center point of the middle lane. If the Ball moves to the right subsection belonging to the user's path, the user feels a leftward attractive force. On the contrary, when the ball passes to the computer player's side, then the feels a rightward repulsive force. This haptic information signals a collaboration opportunity to the user, but if the user does not accommodate the computer player, a physical conflict occurs.

---

[2]41 out of 45 obstacle combinations cause conflicting circumstances.

Experimental Design

A key issue in human-computer interaction is the communication of the affect between parties. Most work on the synthesis of affective cues have focused on the generation of custom tailored facial animations, displays, gaze patterns and auditory signals conveying affective state. Haptics, on the other hand, has been little explored in this context. Moreover, work on affect synthesis has focused mostly on the generation of the traditional set of six basic emotions [16]. With the second experiment, we focus on how haptic and audio-visual cues can be used to convey a specific set of negotiation behaviors that are accompanied by negotiation-related complex affective states. There are different classifications of the negotiation behaviors in the literature (e.g. [17]). However, we preferred to use broader definitions and implemented three negotiation behaviors, namely *concession*, *collaboration*, and *modified tit-for-tat* for the computer player:

Concession   The computer player makes concessions for the benefit of the human. Specifically, the computer player's main goal is to let the Ball collect the coin in the middle lane. This movement allows the human to collect his/her own coin without any compromises. For several conditions that we defined, the computer chooses to concede and help the ball collect the coin in the middle lane. For every obstacle combination, the computer player makes a concession if the value of the coin that the Ball can collect is equal to or higher than its own coin's value. Additionally, the computer player evaluates the difference between the human's benefit and the cost of its concession. The human's benefit is the sum of the values of the coins that the Ball and HIP will collect, whereas the cost of the computer's concession is the value of the coin that it chooses not to collect. If the human's benefit outweighs the computer' cost, then the computer makes a concession. Lastly, the average value of the players' coins is calculated. If this value is less than the value of the coin that the Ball can collect, then the computer player makes a concession. If none of these conditions are met, the computer player ignores the human player, and collects its own coin.

Competition   In competition, each party has its own interests which are conflicting with each other. The game theory literature considers the competitive negotiation as a win-lose type of negotiation [18]. We implemented the competitive computer player to value its interests more than it values those of the human player. In case of conflicts, the computer player tends to collect its own coin to increase its score. However, a result of this persistent, non-cooperative attitude, if both parties try to increase their individual utilities, they can miss some valuable opportunities since none of the parties will consider collecting the coin in the middle lane. The computer chooses to compete in two conditions. First, if the computer's coin is more valuable than the coin in the middle, the computer collects its own coin. Second, the computer player evaluates the benefit of making a concession. It weighs the amount of increase in its earnings relative to the value of the human's coin. Unless the incremental benefit exceeds HIP's earning, the computer player carries on collecting its own coin. If none of these conditions holds, then the computer player accommodates the human and helps the Ball to collect the coin in the middle lane.

Modified Tit-for-Tat   Tit-for-tat is an effective strategy in game theory for the solution of the prisoner's dilemma problem. Tit-for-tat is a cooperative negotiation strategy. Guttman and Maes [18] classifies cooperative negotiation as a decision-making process of resolving a conflict involving two or more parties with non-mutually exclusive goals. We modified the classical tit-for-tat strategy by incorporating some conditions. Initially the computer player starts with a cooperating move. The computer player continues cooperation until the human player defects.[3]

We investigated whether the subjects could differentiate between these negotiation behaviors or not. We also examined the subjects' perceptions on different playing strategies of the computer player. Moreover, we sought an indication of the effectiveness of different sensory modalities on the negotiation process. Finally, we evaluated the performances of the subjects on how effectively they could utilize these negotiation behaviors.

24 subjects (5 female, 19 male) participated in our experiment. Half of these subjects was tested under a visual and haptic feedback (VH) condition, and the remaining half was tested under a visual feedback only (V) condition. In each sensory feedback condition (V and VH), three negotiation behaviors (modes) were tested. There are six combinations for the ordering of three different negotiation behaviors. In order to eliminate the ordering effects, each combination was played by 2 different subjects. The subjects were informed about the existence of three different playing behaviors of the computer player, and were instructed to pay attention the computer's behavior in each mode.

Initially the subjects practiced with a test game (labeled Mode R), in which negotiation behaviors of the computer were randomized so that the subjects did not acquire prior knowledge about the negotiation behaviors. Afterwards, the actual experiment began and subjects played the game once in each behavior of the computer. While playing the game, subjects were not aware of the negotiation behavior the computer player adopted. While they were playing the game, they could only see a reference to the mode of the computer player, (e.g. Mode A, Mode B, or Mode C), on the screen. Finally, all 3 behaviors were displayed to the subjects in succession in order to enable them to compare the behaviours.

Measures

Quantitative Measures   We quantified the subjects' performance in terms of the utility of the game and the average force that the subjects feel through the haptic device. The utility of the game is calculated using the individual scores of the dyads, and the score obtained from the middle lane, which is interpreted as the joint score of the parties.

---

[3]Defection of the human player means that he or she does not accommodate the computer player in keeping the Ball on its path.

The individual and overall utilities of the process for each negotiation behavior and feedback condition is calculated as follows:

$$\text{Individual Utility} = \frac{\text{Achieved Individual Score}}{\text{Maximum Achievable Individual Score}}, \tag{13.1}$$

$$\text{Overall Utility} = \frac{\sum \text{Achieved Individual Scores}}{\text{Maximum Achievable Overall Score}}. \tag{13.2}$$

The average force that the users feel through the haptic device is calculated using the mass-spring-damper system between the interface points and is assumed to be the definitive indicator of the collaboration or the conflict between dyads.

Subjective Evaluation At the end of the experiment, subjects were asked to fill out a short questionnaire regarding their experience and the behavior of the computer player.[4] The subjects answered 15 questions on a 7-point Likert scale, 7 of which were on personal information. A single question asked for user feedback and 7 questions were about the variables directly related to our investigation. Some of the questions were paraphrased, and asked again, but scattered randomly in the questionnaire. In evaluation, the averages of the responses to the questions that fall into the same category are considered. Questions were asked in four categories:

Performance: We asked the subjects to evaluate the computer player's and their own performance (1 question each).

Perceived Sense of Conflict and Collaboration: We asked the subjects whether or not the subjects had a sense of conflict (2 questions) or collaboration (2 questions) with the computer.

Effectiveness of Modalities: We asked the subjects to rate the perceived effectiveness of the three modalities -audio, visual, and haptics—for helping them identify the behaviors of the computer player (1 question).

13.4 Results

13.4.1 Haptic Board Game

This section summarizes the key results of the experiment conducted with the haptic board game.

Quantitative Measures

As mentioned in Sect. 13.3.1, in the haptic board game, task performance is quantified in terms of task completion time, penalized completion time, and the number of

---

[4]For the questionnaire design, we adopted the technique that Basdogan [14] used previously in shared virtual environments.

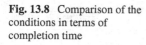

Fig. 13.8 Comparison of the conditions in terms of completion time

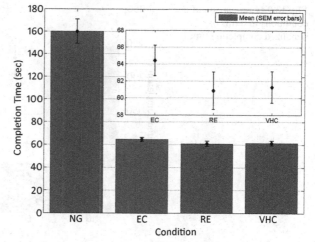

faults during trials. For quantitative analysis, we utilized the data collected during the evaluation session. Even though NG was included in the experiment solely to get the users familiar with the system, the data acquired under NG is also presented for the sake of completeness.

Figure 13.8 shows a comparison of the means of the conditions in terms of task completion time. Since the variance of completion time in no guidance condition is much higher than those in the other conditions, the comparison of the guidance conditions is included as a close-up in the upper right corner of the figure. Wilcoxon Signed-Rank Test indicates a statistically significant difference between NG and all three guidance condition in terms of completion time (p-value < 0.01). We observe that role exchange (RE and VHC) improves completion time when compared to NG and EC. The completion time in EC is observed to be significantly higher than it is in RE (p-value < 0.01), and no significant difference is observed between RE and VHC. Hence we conclude that although adding certain sensory cues (VCH) seems to increase the completion time, this time is still close to that in RE.

Figure 13.9 illustrates the differences between the conditions in terms of penalized completion time. Again, due to the high variability in NG, a separate comparison of the guidance conditions is given as a close-up in the upper right corner of the figure. Penalized completion time in NG is significantly inferior than it is the other conditions (p-value < 0.01), whereas it is the best in RE, followed by VHC and EC. Even though the differences is significant between EC and RE (p-value < 0.05), no other conditions exhibit significant differences among each other.

Finally, Fig. 13.10 suggests that adding sensory cues on top of a role exchange mechanism (VHC condition) slightly increases the errors made by the user during the execution of the task. However, this increase is not statistically significant.

Along with the performance measures, we calculated the energy spent by the user in order to complete the task and the work done by him/her. Figure 13.11 shows the mean values and the standard error of means for each condition. As expected, the users spent the largest amount energy under NG, when no guidance is available.

Fig. 13.9 Comparison of the conditions in terms of penalized completion time

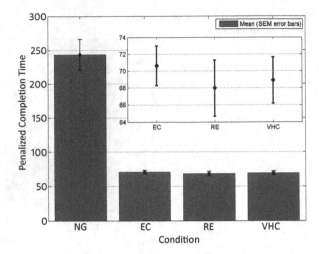

Fig. 13.10 Comparison of the number of faults in each condition

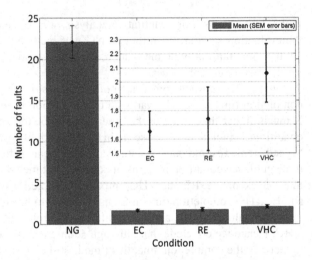

Wilcoxon Signed-Rank Test reveals that the energy requirement under NG is significantly higher than those under the guidance conditions (p-value < 0.001). The guidance conditions display no significant difference, however can be sorted in ascending order of the amount of energy that the users spend as EC, RE, and VHC.

Figure 13.12 illustrates a comparison of the average work done by the user for completing the task under each condition. We observe that the work done by the user towards task completion is maximized in NG. On the contrary, the work done in EC is smaller than those in RE and VHC (p-value < 0.05).

We defined efficiency as the ratio of work done by the user on the ball to the energy (s)he spends. Figure 13.13 displays the average efficiencies of the users for each condition. The efficiency of users is low in NG probably because their control on the ball is not good. Note that the worst completion time and the greatest number of faults are made under this condition. We observe that the efficiency in EC is

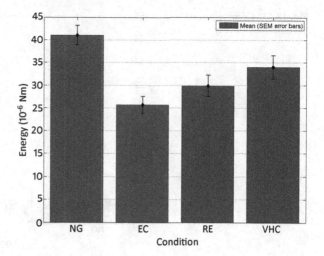

Fig. 13.11 Comparison of the energy spent by the user

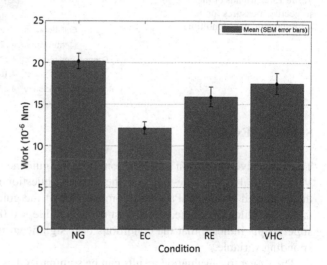

Fig. 13.12 Comparison of the work done by the user on the ball

also as low as it is in NG. On the other hand, the efficiencies under RE and VHC are significantly higher than the efficiencies under NG and EC (p-value < 0.005). We also observed a statistically significant difference between the efficiencies under RE and VHC (p-value < 0.05), RE having the highest efficiency. As a result, we conclude that even though the energy spent by the users is low in EC, the users fail to do work on the ball. Instead they surrender to computer guidance and take less initiative in performing the task, which decreases their efficiency. On the other hand, the efficiencies in RE and VHC are high, indicating that the added effort is effectively used on the task.

Fig. 13.13 Comparison of
the efficiency of user

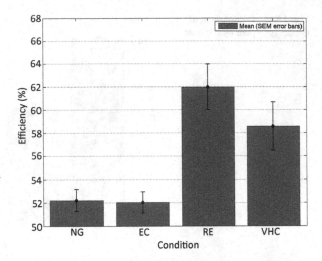

Table 13.2 Means of the
subjective measures for
different guidance conditions

	EC		RE		VHC	
	Mean	SD	Mean	SD	Mean	SD
Collaboration	4.2^a	1.35	4.47^a	1.14	5.01^b	0.86
Interaction	4.12^a	1.17	4.47^a	0.87	4.84^b	0.67
Comfort and pleasure	4.12^a	1.04	4.32^{ab}	0.91	4.62^b	0.93
Trust	4.18^a	1.57	4.57^a	1.47	5.2^b	0.92
Role exchange visibility	3.27^a	1.99	3.87^a	2.03	4.67^b	1.56

Subjective Evaluation

The subjective evaluation was done only for the guidance conditions EC, RE, and
VHC. Table 13.2 lists the means of the subjects' questionnaire responses regarding
the evaluated variables. Pair-wise comparisons of the guidance conditions are obtained by Wilcoxon Signed-Rank Test using p-value < 0.05. Different letters in the
superscripts indicate that the conditions bear significant differences for the corresponding variable.

The subjective evaluation results can be summarized as follows:

- *Collaboration:* Additional sensory cues significantly improve the sense of collaboration during the task when they are displayed to the user to indicate the control
  state.
- *Interaction:* Additional sensory cues significantly improve the interaction level of
  the task.
- *Comfort and pleasure:* When compared to an equal control condition, the users
  find the interface significantly more comfortable, enjoyable, and easier to perform, only if additional sensory cues are provided to them.
- *Trust:* A role exchange scheme lets the users trust in the computer's control during
  collaboration, such that they believe that it will move the ball correctly. This sense
  of trust is significantly higher when additional sensory cues are present.

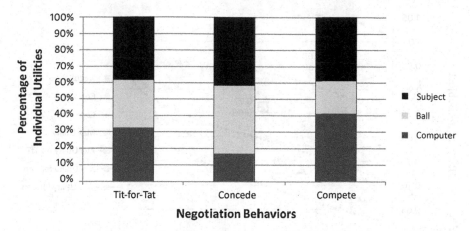

Fig. 13.14 Percentage of the individual utilities while normalized by the overall utility

- *Role exchange visibility:* Additional sensory cues make the role exchange process significantly more visible so that the users can track the current state of the system more successfully when these cues are present.

13.4.2 Haptic Negotiation Game

This section presents the results of the experiment conducted with the haptic negotiation game.

Quantitative Measures

We investigated how humans interact with a computer that executes different negotiation strategies. We looked into whether the users can utilize any of those strategies in the haptic negotiation game or not. We also studied the role of haptic force feedback on these interactions. Hence, we computed the average force values that the users felt, as well as the individual and overall utilities of the process for each negotiation behavior with each feedback condition using (13.1) and (13.2).

We observed differences between the individual utilities of the dyads while the computer player is playing with three different negotiation behaviors. As expected, in concession and competition, a single player's individual utility is favored. For example, the computer player making numerous concessions result in higher individual utility for the human user. Figure 13.14 displays the utility percentages of the human and the computer players. In the figure, the joint utility of the dyads is marked as the ball's utility. Clearly, in concession, the computer player's utility is low since it continuously makes concessions to improve the human player's utility. In competition, the computer player focuses only improving its individual utility,

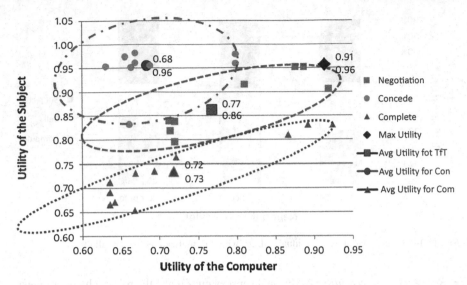

Fig. 13.15 Utility distribution of the subjects, their means and the ellipses showing the clustering of the different negotiation behaviors under VH condition. (Tft: tit-for-tat, Con: Concession, Com: Competition)

hence it impairs the utility of the human player and the joint utility. On the other hand, the tit-for-tat strategy favors the overall utility without sacrificing the individual utilities of the dyads. Hence, we observe a more uniform percentage distribution in both the players' and the joint utilities.

Figure 13.15 displays the distributions of the individual utilities for each of the three negotiation behaviors. The utility values of the negotiation behaviors are clustered as shown in the figure, that clearly shows the distinction of achieved utility values as indicated by the dashed ellipsoids. The average utility value for each negotiation behavior is also shown in Fig. 13.15, at the center of the enclosing ellipses. When we consider the distance between those average utilities to the achievable maximum utility value, we observe that the tit-for-tat strategy is the closest one to that value. This confirms that the dyads were able to utilize this strategy, and can achieve utility values which are closer to the ideal condition.

In order to test the effectiveness of force feedback on a negotiation framework, we calculated the average forces that are felt by the subjects in VH condition as plotted in Fig. 13.16. We interpret the force felt by the user is an indication of the effort (s)he spends. Unsurprisingly, the competitive computer player required the human player to make the greatest effort. With this strategy, the human players faced conflicting situations most of the time, and due to these conflicts, higher force values were generated and fed back to the user. On the other hand, the concessive computer player accommodated the human player, and created fewer conflicts. As a result of this, the user, spending the least effort, felt a small force. Finally, the forces generated with the tit-for-tat strategy fell in between the other two strategies. Hence, we conclude that tit-for-tat provides a trade-off between the effort and the utility.

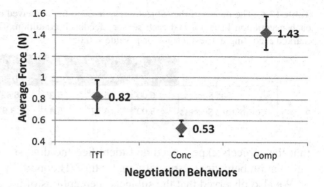

Fig. 13.16 Average forces that the subjects felt through the haptic device for each negotiation behavior. (Tft: tit-for-tat, Conc: Concession, Comp: Competition)

Table 13.3 The means of the subjects' responses to the perceived sense of conflict and collaboration questions for each negotiation behavior and each sensory feedback condition. The significant differences are marked according to t-test results for p-value < 0.05

	V			VH		
	Tft	Con	Com	Tft	Con	Com
Sense of Conflict	3.46[a]	2.92[a]	5.62[b]	4.33[a]	1.38[d]	6.5[b]
Sense of Collaboration	5.04[a]	5.46[b]	3.50[c]	4.54[a]	6.67[d]	1.67[e]

In other words, with a little more effort made by two parties, they can maximize the overall utility of the application. Our analysis is in line with the assumptions presented in the negotiation literature, hence indicate that the negotiation behaviors have been successfully implemented in the haptic negotiation game.

Subjective Evaluation

The subject's responses to (a) self success, (b) computer player's success (c) perceived sense of conflict, (d) perceived sense of collaboration, and (e) effectiveness of modalities are used as variables to evaluate the statistical significance of the differences between negotiation behaviors.

Our results show that the subjects can successfully differentiate between the behaviors of the computer. Table 13.3 shows the means and significant differences between negotiation behaviors under V and VH, obtained by t-test using p-value < 0.05. Significant differences for a variable are indicated by different letters in the superscripts. We observed that all variables exhibit a clear distinction between the negotiation behaviors, when visual and haptic feedback is presented simultaneously (VH). In both V and VH conditions, the subjects were successful at identifying the collaborative and conflicting characteristics of the computer player. We observe that there is a clear distinction for all the behaviors in VH. However, under V, there are cases where significant differences cannot be observed between the negotiation behaviors. For example, the subjects could not differentiate between the tit-for-tat and the concession behavior when evaluating how much the computer player was working against them (p-value < 0.05). These results evidently suggest

Table 13.4 The means of the subjects' responses to the perceived sense of success questions for each negotiation behavior and each sensory feedback condition. The significant differences are marked according to t-test results for p-value < 0.05

	V			VH		
	Tft	Con	Com	Tft	Con	Com
Self Success	6.00^a	6.42^b	5.25^c	6.42^{ad}	6.83^d	5.58^c
Computer's Success	5.00^a	4.17^b	5.42^a	3.92^a	2.83^b	5.17^a

that the subjects experienced and identified the diversity of the computer player's negotiation behaviors, especially under the VH condition.

We also observed that the subjects' perceptions of the computer player's performance were not statistically significant between the tit-for-tat and the competitive behavior (p-value < 0.05). Likewise, the subjects could not perceive a significant difference between their own performances in tit-for-tat and the concessive behavior under VH (see Table 13.4).

Finally, we examined how effectively the three modalities support the subjects to differentiate between the behaviors of the computer player. In VH condition, the effectiveness of haptic feedback is found to be superior to the audio and the visual channels. Subjects rated the effectiveness of the haptic channel with an average score as high as 6.33, whereas they rated the visual and audio channels respectively with 5.25 and 2.25, respectively. The difference between these scores display statistical significance between the haptic and the other two feedback modalities (p-value < 0.05) according to Mann-Whitney U-Test. Hence, the haptic feedback proves to be an effective initiator for the subjects to comprehend the cues of their negotiation with the computer player.

13.5 Conclusions

Haptic human-computer cooperation has applications in entertainment, teleoperation, and training domains. As such applications are becoming common, a need to develop more sophisticated interaction schemes between a human operator and a computer emerged. The proposed research lays foundations for a natural and human-like cooperation scheme by examining the interaction dynamics between the parties. This chapter reports the results of two experiments conducted to examine the possible usage of a dynamic role exchange mechanism and a haptic negotiation framework to enable and improve human-computer cooperation in a virtual environment.

The proposed negotiation model can be used to realize decision making and negotiation. This model is tested through two cooperative games to display collaboration and conflicts between the parties. We developed a set of scales for objective and subjective evaluation of the users in these games. The results suggest that the negotiation model can be used (1) as a role exchange mechanism and (2) to enable negotiation between the parties and assign different behaviors to the computer. This model, when used as a role exchange mechanism in a dynamic task, increases the task performance and the efficiency of the users. It also provides a natural and seamless interaction between the human and the computer, improving collaborative and interactive aspects of the task. The same mechanism can be used to display negotiation related affective states. We showed that the addition of haptic cues provides a statistically significant increase in the human-recognition of these machine-displayed affective cues. If the same affective cues are also recognized by the computer, then more user-specific cooperation schemes compensating for the weaknesses of the human operator can be developed. Also, we showed that, certain negotiation strategies, such as tit-for-tat, generate maximum combined utility for the negotiating parties while providing an excellent balance between the work done by the user and the joint utility.

As a future work, we will explore the utility of statistical learning models to investigate certain classifications for the human user and extract behaviours for the computer that works best for each human class. These ideas can further be extended to realize the haptic version of the Turing test, where the computer will be programmed to mimic the haptic interactions between two human beings.

References

1. Sheridan, T.B.: Telerobotics, Automation, and Human Supervisory Control, p. 393. MIT Press, Cambridge (1992)
2. O'Malley, M.K., Gupta, A., Gen, M., Li, Y.: Shared control in haptic systems for performance enhancement and training. J. Dyn. Syst. Meas. Control 128(1), 75–85 (2006). doi:10.1115/1.2168160
3. Crandall, J.W., Goodrich, M.A.: Experiments in adjustable autonomy. In: IEEE International Conference on Systems, Man, and Cybernetics, vol. 3, pp. 1624–1629 (2001). doi:10.1109/ICSMC.2001.973517
4. Sierhuis, M., Bradshaw, J.M., Acquisti, A., van Hoof, R., Jeffers, R., Uszok, A.: Human-agent teamwork and adjustable autonomy in practice. In: Proceedings of the Seventh International Symposium on Artificial Intelligence, Robotics and Automation in Space (I-SAIRAS) (2003)
5. Nudehi, S.S., Mukherjee, R., Ghodoussi, M.: A shared-control approach to haptic interface design for minimally invasive telesurgical training. IEEE Trans. Control Syst. Technol. 13(4), 588–592 (2005). doi:10.1109/TCST.2004.843131
6. Reed, K.B., Peshkin, M.A.: Physical collaboration of human-human and human-robot teams. IEEE Trans. Haptics 1(2), 108–120 (2008). doi:10.1109/TOH.2008.13
7. Stefanov, N., Peer, A., Buss, M.: Role determination in human-human interaction. In: WHC '09: Proceedings of the World Haptics 2009—Third Joint EuroHaptics conference and Symposium on Haptic Interfaces for Virtual Environment and Teleoperator Systems, Washington, DC, USA, pp. 51–56. IEEE Computer Society, Los Alamitos (2009). doi:10.1109/WHC.2009.4810846
8. Groten, R., Feth, D., Peer, A., Buss, M.: Shared decision making in a collaborative task with reciprocal haptic feedback—an efficiency-analysis. In: IEEE International Conference on Robotics and Automation (ICRA), pp. 1834–1839 (2010). doi:10.1109/ROBOT.2010.5509906
9. Groten, R., Feth, D., Klatzky, R., Peer, A., Buss, M.: Efficiency analysis in a collaborative task with reciprocal haptic feedback. In: IROS'09: Proceedings of the 2009 IEEE/RSJ International Conference on Intelligent Robots and Systems, Piscataway, NJ, USA, pp. 461–466. IEEE Press, New York (2009)

10. Evrard, P., Kheddar, A.: Homotopy switching model for dyad haptic interaction in physical collaborative tasks. In: WHC '09: Proceedings of the World Haptics 2009—Third Joint EuroHaptics conference and Symposium on Haptic Interfaces for Virtual Environment and Teleoperator Systems, Washington, DC, USA, pp. 45–50. IEEE Computer Society, Los Alamitos (2009). doi:10.1109/WHC.2009.4810879
11. Oguz, S.O., Kucukyilmaz, A., Sezgin, T.M., Basdogan, C.: Haptic negotiation and role exchange for collaboration in virtual environments. In: IEEE Haptics Symposium, pp. 371–378 (2010). doi:10.1109/HAPTIC.2010.5444628
12. Kucukyilmaz, A., Sezgin, T.M., Basdogan, C.: Conveying intentions through haptics in human-computer collaboration. In: IEEE World Haptics Conference, pp. 421–426 (2011). doi:10.1109/WHC.2011.5945523
13. Oguz, S.O., Sezgin, T.M., Basdogan, C.: Supporting negotiation behavior in haptics-enabled human-computer interfaces. IEEE Trans. Haptics (2011, in review)
14. Basdogan, C., Ho, C.-H., Srinivasan, M.A., Slater, M.: An experimental study on the role of touch in shared virtual environments. ACM Trans. Comput.-Hum. Interact. 7(4), 443–460 (2000). doi:10.1145/365058.365082
15. Dorf, R.C., Bishop, R.H.: Modern Control Systems. Prentice-Hall, Upper Saddle River (2000). ISBN:0130306606
16. Ekman, P.: Basic emotions. In: Dalgleish, T., Power, M.J. (eds.) Handbook of Cognition and Emotion, pp. 45–60. Wiley, New York (1999)
17. Shell, G.: Bargaining for Advantage: Negotiation Strategies for Reasonable People. Penguin Books, New York (1999)
18. Guttman, R.H., Maes, P.: Cooperative Vs. Competitive multi-agent negotiations in retail electronic commerce. In: Klusch, M., Weiss, G. (eds.) Cooperative Information Agents II Learning, Mobility and Electronic Commerce for Information Discovery on the Internet. Lecture Notes in Computer Science, vol. 1435, pp. 135–147. Springer, Berlin (1998)

Index